Parasitology

Parasitology

Elen Downey

Callisto Reference

www.callistoreference.com

Callisto Reference,
118-35 Queens Blvd., Suite 400,
Forest Hills, NY 11375, USA

Visit us on the World Wide Web at:
www.callistoreference.com

ISBN: 978-1-64116-549-5 (Hardback)

Cataloging-in-Publication Data

Parasitology / Elen Downey.
p. cm.
Includes bibliographical references and index.
ISBN 978-1-64116-549-5
1. Parasitology. 2. Biology. I. Downey, Elen.
QL757 .P37 2022
591.785 7--dc23

Table of Contents

Preface

The purpose of this book is to help students understand the fundamental concepts of this discipline. It is designed to motivate students to learn and prosper. I am grateful for the support of my colleagues. I would also like to acknowledge the encouragement of my family.

The study of parasites and their hosts is referred to as parasitology. It also includes the study of the relationship between them. It is a biological field that draws on techniques from various fields such as bioinformatics, biochemistry, cell biology, genetics, ecology, immunology, molecular biology and evolution. Some of the branches of parasitology are medical parasitology, veterinary parasitology and structural parasitology. Medical parasitology deals with the parasites that infect humans and the diseases caused by them. It also focuses on various methods used to diagnose, prevent, control and treat such diseases. The study of parasites that leads to economic losses in agriculture and aquaculture operations infecting the companion animals falls under veterinary parasitology. This book explores all the important aspects of parasitology in the present day scenario. It aims to shed light on some of the unexplored aspects of this field. This book will serve as a valuable source of reference for those interested in this field.

A foreword for all the chapters is provided below:

Chapter - Introduction

The discipline which studies parasites, their hosts and the relationship between them is known as parasitology. This is an introductory chapter which will introduce briefly about the major types of parasites such as endoparasites, ectoparasites and complex lifecycle parasites.

Chapter - Branches of Parasitology

The three main branches of parasitology are medical parasitology, veterinary parasitology and paleoparasitology. These branches deal with the parasites which infect humans, parasites which infect animals and parasites from the past, respectively. This chapter has been carefully written to provide an easy understanding of the varied facets related to these branches of parasitology.

Chapter - Protozoa

Protozoa are the eukaryotes that feed on organic matter such as other microorganisms, organic tissues and debris. The topics elaborated in this chapter will help in gaining a better perspective about the different types of protozoa such as leishmania, balantidum coli and entamoeba.

Chapter - Parasitic Worms

Parasitic worms are macroparasites such as zoonotic hookworms, threadworms, filarial worms, giardia lambia, cestoda, human whipworm and ascaris lumbricoides. They can be either endoparasitic or ectoparasitic. This chapter has been carefully written to provide an easy understanding of these types of parasitic worms.

Chapter - Parasitic Arthropods

The arthropods are the invertebrate animal having an exoskeleton, paired jointed appendages and a segmented body. Some of the common parasitic arthropods are fleas, demodex and ticks. The topics elaborated in this chapter will help in gaining a better perspective about these types of arthropod parasites.

Chapter - Intracellular Parasites

The microparasites that can grow and reproduce inside the cells of a host are known as intracellular parasites. Cystoisospora belli, coccidian and toxoplasma are some examples of intracellular parasites. This chapter closely examines these key types of intracellular parasites to provide an extensive understanding of the subject.

Chapter - Parasites in Plants

Parasitic plants are the plants that obtain their nutritional requirement from other living plants. Dodder, nuytsia, orobanche minor, viscum album and orobanche unifloraare are some of the common examples of parasitic plants. This chapter has been carefully written to provide an easy understanding of these parasitic plants.

Elen Downey

1
Introduction

The discipline which studies parasites, their hosts and the relationship between them is known as parasitology. This is an introductory chapter which will introduce briefly about the major types of parasites such as endoparasites, ectoparasites and complex lifecycle parasites.

Parasite

A parasite is an organism that lives in another organism, called the host, and often harms it. It depends on its host for survival.

Without a host, a parasite cannot live, grow and multiply. For this reason, it rarely kills the host, but it can spread diseases, and some of these can be fatal.

Parasites, unlike predators, are usually much smaller than their host and they reproduce at a faster rate.

- Parasites live on or in other organisms and thrive to the detriment of their host.
- Many different parasites can affect humans, and they can pass on diseases such as malaria and trichomoniasis.
- Ensuring that food is fully cooked, using insect repellant, and following good hand hygiene rules can reduce the risk of getting parasites.

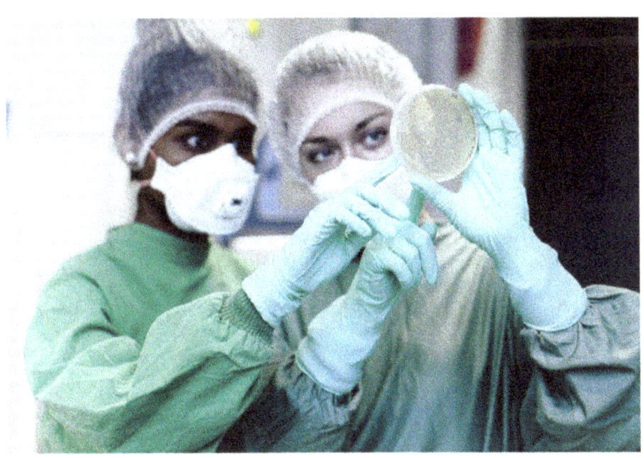

Parasites range from microscopic in size to over 30 meters in length.

A parasite is an organism that lives within or on a host. The host is another organism. The parasite uses the host's resources to fuel its life cycle. It uses the host's resources to maintain itself. Parasites vary widely. Around 70 percent are not visible to the human eye, such as the malarial parasite, but some worm parasites can reach over 30 meters in length. Parasites are not a disease, but they can spread diseases. Different parasites have different effects.

Endoparasite

These live inside the host. They include heartworm, tapeworm, and flatworms. An intercellular parasite lives in the spaces within the host's body, within the host's cells. They include bacteria and viruses.

Endoparasites rely on a third organism, known as the vector, or carrier. The vector transmits the endoparasite to the host. The mosquito is a vector for many parasites, including the protozoan known as Plasmodium, which causes malaria.

Epiparasite

These feed on other parasites in a relationship known as hyperparasitism. A flea lives on a dog, but the flea may have a protozoan in its digestive tract. The protozoan is the hyperparasite.

Types

There are three main types of parasites:

- Protozoa: Examples include the single-celled organism known as Plasmodium. A protozoa can only multiply, or divide, within the host.
- Helminths: These are worm parasites. Schistosomiasis is caused by a helminth. Other examples include roundworm, pinworm, trichina spiralis, tapeworm, and fluke.
- Ectoparasites: These live on, rather than in their hosts. They include lice and fleas.

Symptoms

There are many types of parasite, and symptoms can vary widely. Sometimes these may resemble the symptoms of other conditions, such as a hormone deficiency, pneumonia, or food poisoning.

Symptoms that might occur include:

- Skin bumps or rashes;
- Weight loss, increased appetite, or both;
- Abdominal pain, diarrhea, and vomiting;
- Sleeping problems;
- Anemia;

- Aches and pains;
- Allergies;
- Weakness and general feeling unwell;
- Fever.

However, parasites can pass on a wide variety of conditions, so symptoms are hard to predict.

Often there are no symptoms, or symptoms appear long after infection, but the parasite can still be transmitted to another person, who may develop symptoms.

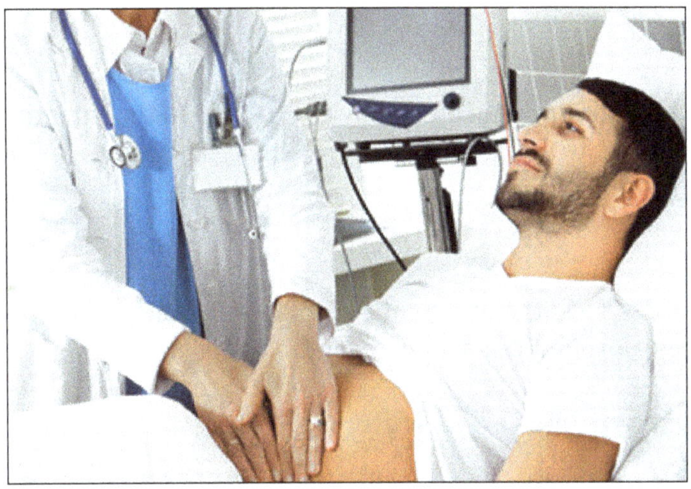

Some parasite-related problems, such as giardiasis and amebic dysentery, can cause abdominal pain.

Human Parasites

Many types of parasites can affect humans. Here are some examples of parasites and the diseases they can cause.

Acanthamoebiasis

This tiny ameba can affect the eye, the skin, and the brain. It exists all over the world in water and soil. Individuals can become infected if they clean contact lenses with tap water.

Babesiosis

This disease that comes from parasites that are spread by ticks. It affects the red blood cells. The risk is highest in summer in the Northeast and upper Midwest of the United States.

Balantidiasis

This is passed on by Balatidium coli, a single-cell parasite that usually infects pigs but can, in rare cases, cause intestinal infection in humans. It can be spread through direct contact with pigs or by drinking contaminated water, usually in tropical regions.

Blastocystosis

This affects the intestines. The blastocystis enters humans through the fecal-oral route. A person can get it by eating food or drink contaminated with human or animal feces where the parasite is present.

Coccidiosis

This affects the intestines. Coccidia is passed on through the fecal-oral route. It is found around the world. It can also affect dogs and cats, but these are different kinds. Dogs, cats, and humans cannot normally infect each other.

Amoebiasis

This is caused by the parasite Entamoeba histolytica. It affects the intestines. It is more likely in tropical regions and in areas with high population density and poor sanitation. It is transmitted through the fecal-oral route.

Giardiasis

Giardia, or "beaver fever" affects the lumen of the small intestine. If humans ingest food or water contaminated with feces, dormant cysts may infect the body.

Isosporiasis or Cystosporiasis

This disease is caused by the Cystoisospora belli, previously known as Isospora belli. It affects the epithelial cells of the small intestine. It exists worldwide and is both treatable and preventable. It is passed on through the fecal-oral route.

Leishmaniasis

This is a disease that is passed on by parasites of the Leishmania family. It can affect the skin, the viscera, or the mucous membranes of the nose, mouth, and throat. It can be fatal. The parasite is transmitted by types of sandflies.

Primary Amoebic Meningoencephalitis (PAM)

This is passed on through a free-living ameba known as Naegleria fowleri. It affects the brain and the nervous system, and it is nearly always fatal within 1 to 18 days. It is transmitted through breathing in contaminated soil, swimming pools, and contaminated water, but not from drinking water.

Malaria

Different types of plasmodium affect the red blood cells. It exists in tropical regions and is transmitted by the Anopheles mosquito.

Rhinosporidiosis

This is caused by Rhinosporidium seeberi. It mainly affects the mucous of the nose, conjunctiva, and urethra. It is more common in India and Sri Lanka but can occur elsewhere. Polyps result in nasal masses that need to be removed through surgery. Bathing in common ponds can expose the nasal mucous to the parasite.

Toxoplasmosis

This is a parasitic pneumonia caused by the parasite Toxoplasma gondii. It affects the liver, heart, eyes and brain. It occurs worldwide. People can become infected after ingesting raw or undercooked pork, lamb, goat, or milk, or though contact with food or soil that is contaminated with cat feces.

A person with a healthy immune system will not usually have symptoms, but it can pose a risk during pregnancy and for those with a weakened immune system.

Trichomoniasis

Also known as "trich" this is a sexually transmitted infection (STI) caused by the parasite Trichomonas vaginalis. It affects the female urogenital tract. It can exist in males, but usually without symptoms.

Trypanomiasis (Sleeping Sickness)

This is passed on when the tetse fly transmits a parasite of the Trypanosoma family. It affects the central nervous system, blood, and lymph. It leads to changes in sleep behavior, among other symptoms, and it is considered fatal without treatment. It can cross the placenta and infect a fetus during pregnancy.

Chagas Disease

This affects the blood, muscle, nerves, heart, esophagus and colon. It is transmitted through an insect bite. Over 300,000 people in the U.S. have the parasite that can lead to this disease.

Worms

Worms, or helminth organisms, can affect humans and animals.

- Anisakiasis: This is caused by worms that can invade the intestines or the stomach wall. The worms are passed on through contaminated fresh or undercooked fish and squid.
- Roundworm: Ascariasis, or a roundworm infection, does not usually cause symptoms, but the worm may be visible in feces. It enters the body through consuming contaminated food or drink.
- Raccoon roundworm: Baylisascaris is passed on through raccoon stools. It can affect the brain, lungs, liver, and intestines. It occurs in North America. People are advised not to keep raccoons as pets for this reason.
- Clonorchiasis: Also known as Chinese liver fluke disease, this affects the gall bladder. Humans can become infected after ingesting raw or poorly processed or preserved freshwater fish.

- Dioctophyme renalis infection: The giant kidney worm can move through the wall of the stomach to the liver and eventually the kidney. Humans can become infected after eating the eggs of the parasite in raw or undercooked freshwater fish.
- Diphyllobothriasis tapeworm: This affects the intestines and blood. Humans can become infected after eating raw fish that live wholly or partly in fresh water. Prevalence has increased in some parts of the developed world, possibly due to the growing popularity of sushi, salted fillets, ceviche, and other raw-fish dishes.
- Guinea worm: This affects subcutaneous tissues and muscle and causes blisters and ulcers. The worm may be visible in the blister. As the worms are shed or removed, they enter the soil or water, and are passed on from there.
- Hookworm: These can cause intestinal disease. They lay their eggs in soil and the larvae can penetrate the skin of humans. Early symptoms include itching and a rash. They are most common in damp places with poor sanitation.
- Hymenolepiasis: Humans can become infected by ingesting material contaminated by rodents, cockroaches, mealworms, and flour beetles.
- Echinococcosis tapeworm: Cystic echinococcosis can lead to cysts in the liver and lungs, and alveolar echinococcosis can cause a tumor in the liver. Humans can be infected after eating foods contaminated by the feces of an infected animal, or from direct contact with an animal.
- Enterobiasis pinworm: A pinworm, or threadworm, Enterobius vermicularis can live in the colon and rectum of humans. The worm lays eggs around the anus while a person sleeps, leading to itching. It spreads through the oral-fecal route.
- Fasciolosis liver fluke: This affects the gall bladder and liver. It is common in countries where cattle or sheep are reared, but rare in the U.S. It can affect the liver and the bile ducts and it causes gastrointestinal symptoms. It passes from one mammal to another through snails. A person may get it from eating watercress, for example.
- Fasciolopsiasis intestinal fluke: This affects the intestines. It can also transmitted when consuming contaminated water plants or water.
- Gnathostomiasis: This causes swellings under the skin, and occasionally affects the liver, the eyes, and the nervous system. It is rare, but it can be fatal. It occurs in Southeast Asia. It is transmitted by eating freshwater fish, pigs, snails, frogs, and chicken.
- Loa loa filariasis: Also known as loaisis, this is caused by the Loa loa worm, or African eye worm. It causes itchy swellings on the body. It occurs mainly in Central and West Africa and is transmitted through deerfly bites.
- Mansonellosis: This is passed on through the bites of midges or blackflies. It affects the layers under the surface of the skin, but it can enter the blood. It can lead to angioedema, swellings, skin rash, fever, and joint problems. It is present in Africa and Cental America.
- River blindness: Caused by a worm known as Onchocerca volvulus, this affects the eyes, skin, and other body tissues. It is found near fast flowing water. It is transmitted through the bite of a blackfly. It occurs in South America, but 90 percent of cases are in Africa.

- Lung fluke: Also known as paragonimiasis, this affects the lungs, causing symptoms similar to those of tuberculosis (TB). However, it can reach the central nervous system, leading to meningitis. It is transmitted when eating undercooked or raw freshwater crabs, crayfishes, and other crustaceans. It is most common in parts of Asia.
- Schistosomiasis, bilharzia, or snail fever: There are different types of schistosomiasis. They can affect the skin and internal organs. It results from exposure to fresh water that has snails in it that are infected with the blood fluke, or trematode worm. The worms are not found in the U.S. but they are common worldwide.
- Sparganosis: Humans can become infected if they eat foods tainted with dog or cat feces that contains the larvae of a tapeworm of the Spirometra family. It can lead to a migrating abscess under the skin. It is rare.
- Strongyloidiasis: This can lead to severe and possibly fatal immunodeficiency. The parasite penetrates through the skin and affects the lungs, skin, and intestines. It is passed on through direct contact with contaminated soil. It most occurs in tropical and subtropical regions.
- Beef and pork tapeworms: Taeniasis is caused by tapeworms of the taenia family. They affect the intestines. They are passed on by eating undercooked beef or pork.
- Toxocariasis: A roundworm transmits this infection from animals to humans. It affects the eyes, brain, and liver. It is caused by accidentally swallowing the eggs of the parasite, for example, when young children play with soil. Nearly 14 percent of people in the U.S. have antibodies, suggesting that millions have been exposed. Most never have symptoms.
- Trichinosis: This is caused by the roundworm of the Trichinella family. Infection can lead to intestinal symptoms, fever, and muscle aches. It is passed on by eating undercooked meat.
- Whipworm: Also known as trichuriasis, whipworms live in the large intestine. Eggs are passed in feces. It is common all over the world. Humans can become infected when ingesting the eggs, for example on unwashed fruit or vegetables.
- Elephantiasis lymphatic filariasis: This is transmitted through mosquito bites. The adult worms live in the lymph system. Infection can lead to lyphedema and elephantiasis, in which swelling can cause disfigurement and disability. In the Americas, it is passed on by the Culex quinquefasciatus mosquito.

Ringworm is sometimes mistaken for a worm, but it is not a worm. It is a fungal infection.

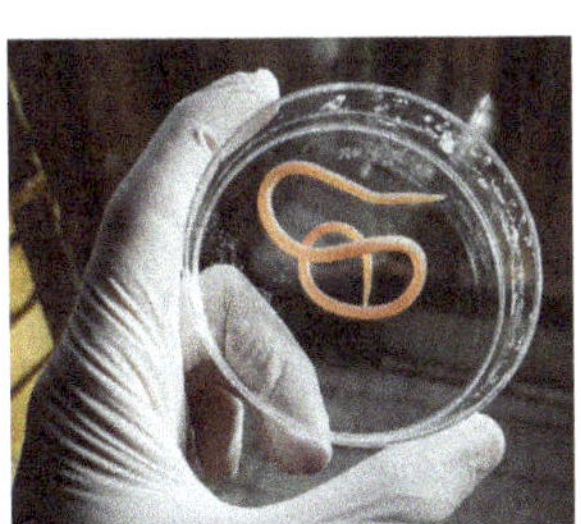

Roundworms can be passed on by raccoons.

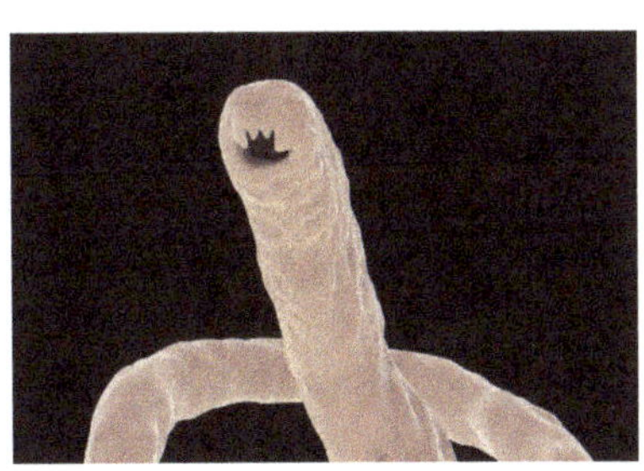

Hookworms can cause intestinal disease.

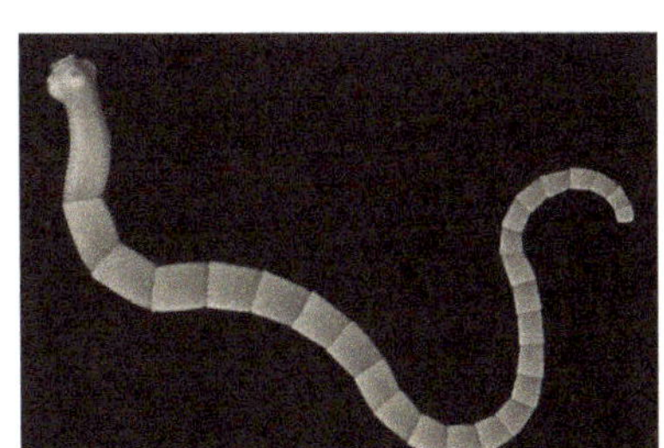

Different types of tapeworm can affect the intestines, the liver, or the lungs.

Parasitism

Parasitism is a type of symbiotic relationship, or long-term relationship between two species, where one member, the parasite, gains benefits that come at the expense of the host member. The word parasite comes from the Latin form of the Greek word (parasitos), meaning "one who eats at the table of another".

Types of Parasitism

There are many types of parasitism, and parasites can belong to multiple classifications based on their size, characteristics, and relationship with the host.

Obligate Parasitism

Obligate parasites are completely dependent on the host in order to complete their life cycle. Over time, they have evolved so that they can no longer exist without the existence of the host. This means that they usually will not cause serious harm to the host, as the parasites need the host to survive so that they can survive, unless the host's death is needed for the transmission of the parasite. Obligate parasitism can be found in many different types of organisms, like plants, animals, fungi, bacteria, and viruses. Head lice are obligate parasites; if removed from the human scalp, they will soon die.

Facultative Parasitism

Facultative parasites do not rely on the host in order to complete their life cycle; they can survive without the host, and only sometimes perform parasitic activities. Certain plants, fungi, animals, and microbes can be facultative parasites. A specific example is the nematode species Strongyloides stercoralis. This is a type of roundworm that can cause the disease strongyloidiasis when it infects humans, but it can also be found free-living.

Ectoparasitism, Endoparasitism and Mesoparasitism

Ectoparasites are parasites that live on the outside of the host's body, such as lice and ticks. Endoparasites, like nematodes and hookworms, live inside the host. Mesoparasites enter the host's external openings, such as the outer ear or the cloaca.

Macroparasitism versus Microparasitism

Macroparasites are parasites that are large enough to be seen with the naked eye. Microparasites are too small to be seen and must be viewed under a microscope. They are generally unicellular, such as protozoa.

Necrotrophic versus Biotrophic

Necrotrophic parasites, also called parasitoids, essentially eat part of the host organism's tissue until it dies from the loss of tissue or from nutrient loss. Biotrophic parasites do not do severe enough damage to kill their host; they need to keep the host alive because they can't survive in a dead one.

Monogenic versus Digenetic

Monogenic parasites complete their life cycle in only one individual host. Digenetic parasites need more than one host to complete their life cycle. Plasmodium vivax, the protozoa that carries malaria, is digenetic. In order to complete its life cycle, it must be a parasite of both people and mosquitos.

Epiparasitism

An epiparasite is a parasite that parasitizes another organism that is also a parasite. Epiparasites are also called hyperparasites or secondary parasites. One example would be a protozoan living in a flea that is living on a dog.

Social Parasitism

Social parasites take advantage of social insects like ants, bees, and termites. They may use mimicry to invade the hive. Some bumblebees invade the hives of other species of bees, making that species raise the parasite's young. One ant species, Tetramorium inquilinum, is a parasite that spends its entire life on the back of other species of ants, essentially making the host species its slaves. This parasite species has gained benefits such as food and transportation, but from this extreme form of parasitism the ants have evolved to be so weak that if they fall off their host, they will not be able to crawl back on, and die.

Brood Parasitism

Brood parasitism involves the raising of young. Bird species that practice brood parasitism, including cowbirds and cuckoos, lay their eggs in another species' nest instead of building their own nests. This is a form of parasitism because the species who lay their eggs in other nests gain benefits (they don't have to spend energy raising young) while the other species are harmed (they do have to use energy to raise young, and it is not their genetic material). Sometimes, the parasite species will even kick the other species' eggs out of the nest, forcing the host to raise only the parasite's young. Brood parasitism can also occur in fish. It is a type of kleptoparasitism, which involves directly or indirectly taking food from the host; in this case, food that could have gone toward the host species goes to the parasite species instead.

This image shows brood parasitism. A brown-headed cowbird has laid its speckled egg in the nest of an Eastern phoebe.

Examples of Parasitism

Over half of all organisms on Earth have a parasitic phase at some point in their life cycle, so there are many examples of parasitism besides the ones already mentioned and the ones listed below.

In Humans

Over 100 different types of organisms can parasitize humans including fungi, leeches, lice, ticks, mites, tapeworms, protozoa, viruses, and helminths. Helminths are worms that can live inside the intestines and can reach meters in length. They can cause a variety of problems such as malnutrition, jaundice, diarrhea, and even in severe cases, death. However, they can be treated with anti-parasitic medication. All infectious diseases, including the common cold, result from organisms that parasitize humans, such as viruses and bacteria. Many of the organisms that parasitize humans can also parasitize other mammals and birds.

In Plants

Aphids are small green insects that parasitize plants by eating their sap. Many types of fungi can also attack plants and can spoil wheat, fruit, and vegetables. Some plants are parasitic themselves. In angiosperms (flowering plants), parasitism has evolved at least 12 separate times, and 4100 species (about 1%) of angiosperms are parasitic. Parasitic plants have haustoria, which are modified roots which connect to the host plant's xylem and/or phloem and drain it of water and nutrients. Some plants parasitize mycorrhizal fungi. This often happens when a plant species has evolved to no longer produce chlorophyll. Since it can no longer photosynthesize, it must gain nutrients for energy in other ways.

In Insects

Entomophagous parasites are insects that parasitize other insects. Usually these parasites attack larva, or young insects. Some insects deposit their eggs within the body of another insect species' larva; when the eggs hatch, the parasitic young kill and eat the larva, gaining nutrients from it. Sometimes, the parent parasite paralyzes a host which is then fed on by the young. This occurs commonly in wasps such as Ampulex compressa, whose young eat paralyzed cockroaches that have been stung by the parent. Other wasps like Ropalidia romandi burrow into the abdomen of their host and then live there. They do not kill their host, but can change its appearance and behavior, and even make it sterile. Parasitism is extremely common in insects. In fact, almost all species of insects are attacked by at least one type of insect parasite.

In Fish

There are many organisms that parasitize fish, and sometimes different populations of the same species of fish living in the same region can be told apart because they have different characteristic parasites. Some parasites, such as copepods (small crustaceans), nematodes, and leeches. attach to the fish's gills and live there. Cymothoa exigua is an isopod (another type of small crustacean) that parasitizes fish. It enters a fish's mouth and eventually severs the fish's tongue. Then, the isopod itself lives where the tongue was, and becomes the new tongue. The host fish can still eat, and will survive with an isopod in its mouth, but the isopod consumes a small amount of the fish's blood and mucus while living there. Cleaner fish like bluestreak cleaner wrasses remove dead skin and parasites from other fish, including large predatory fish that would otherwise eat them. Fish parasites in can be a concern to human health when people eat foods that contain uncooked fish, such as sushi, because the parasites in these fish can also infect humans. However, infection through eating uncooked fish is relatively rare in the developed world, and some raw fish is frozen overnight to prevent infections.

Endoparasites

Endoparasites live inside the body of the host. This group includes helminths, trematodes and cestodes. Endoparasites are two groups of parasites: intercellular and intracellular parasites. Intercellular parasites live in spaces within the host e.g. the alimentary canal, whereas intracellular parasites live in cells within the host e.g. erythrocytes. Intracellular parasites typically rely on a third organism, a vector, to transmit the parasite between hosts. Rather than requiring adaptations to penetrate the host, as ectoparasites do, endoparasites are in a nutrient-rich location so they instead have adaptations to maximise nutrient absorption. Endoparasites have a readily available and renewable supply of nutrients inside the host, which in some cases is pre-digested by the host, so mechanisms of nutrient absorption across their body surface is a common feature. As part of their life cycle strategy, endoparasites must also be able to transmit from within the host body and survive the hostile environment within the host. Only by achieving this can they benefit from acquiring nutrition in this way.

Endoparasites have various anatomical and biochemical adaptations, typically at the host-parasite interface, to maximise nutrient acquisition. One such adaptation is the tegument, a metabolically active external cover that plays an important role in nutrient extraction from the host. The parasite tegument is permeable to various organic solutes and has transporters for the facilitated or active uptake of nutrients. Various studies have attempted to characterise these transporters in a number of parasites e.g. the amino acid transporter molecules in protozoa. Cestodes do not have a gut so the tegument is therefore critical for nutrient uptake. In cestodes the tegument is highly efficient with spine-like microtriches, similar to microvilli, to increase the surface area available for nutrient acquisition. In many parasites the tegument structure has folds or microvilli to maximise the surface area available for diffusion and uptake of nutrients. The tegument also commonly has additional organelles and features with important functions in metabolism including the glycocalyx. The glycocalyx is a carbohydrate-rich layer that enhances nutrient absorption and secretes enzymes to aid primary digestion.

Another important adaptation of endoparasites is the gut, which digests host macromolecules into soluble utilisable products. This feature is particularly important for endoparasites that are not in the alimentary canal, because nutrients are not pre-digested by the host. The gut lining typically has a layer of endodermal cells that secrete proteolytic enzymes to aid digestion. Some endoparasites have both a gut and anus, some lack an anus and some have neither—i.e., those that live in the alimentary canal, which instead diffuse pre-digested host nutrients across their body surface. The relative importance of the tegument, gut and other adaptations involved in nutrient acquisition varies between different endoparasitic species.

Tapeworms

Tapeworms are endoparasites with numerous adaptations to enhance parasitic nutrition. Tapeworms live in the small intestine of humans, providing an ideal location to access a readily available, rich source of pre-digested nutrients. Since nutrients in the small intestine are plentiful and pre-digested by the host, tapeworms do not require a gut and instead have adaptations to maximise nutrient absorption. Tapeworms have a tegument that they use to absorb nutrients directly from the host small intestine by diffusion. They also have anatomical adaptations in the form of a scolex with hookers and suckers that they use to attach to the host small intestine wall, preventing the tapeworm from being egested following peristalsis. Tapeworms have a flattened body with microtriches to maximise the surface area available for nutrient absorption and they additionally have various transporter molecules. Tapeworms have to compete with the host epithelium for nutrients, so it is essential that they compete more efficiently for nutrients. They also secrete enzymes to enhance host digestive enzymes e.g. pancreatic α-amylase.

Schistosomes

Schistosomes, another type of endoparasite, also live inside the body of the host but instead these parasites acquire their nutrients from host blood. Schistosomes are in direct contact with host blood, a rich source of amino acids, and they therefore do not require penetrative structures to reach host nutrients. Schistosomes take blood up through the negative pressure created by muscle contractions of their sucker and oesophagus. They obtain amino acids from host blood through a mechanism of haemoglobin degradation, which remains unresolved but is suggested to involve a series of proteases. Mechanisms to overcome blood clotting are also employed. Various studies have attempted to characterise the components of the schistosome tegument, including transporter molecules that may be involved in nutrient uptake. Such transporter molecules include Schistosome Alkaline Phosphatase (SmAP) and cathepsin B, which may be important in nutrient acquisition.

Malaria

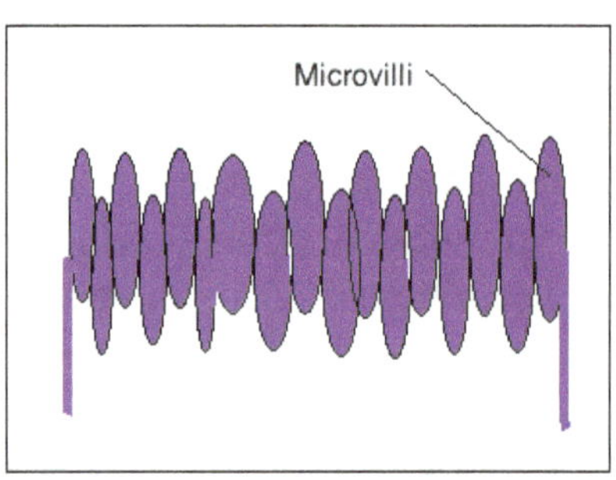

Microvilli structure, which increases the surface area available for nutrient uptake in endoparasites.

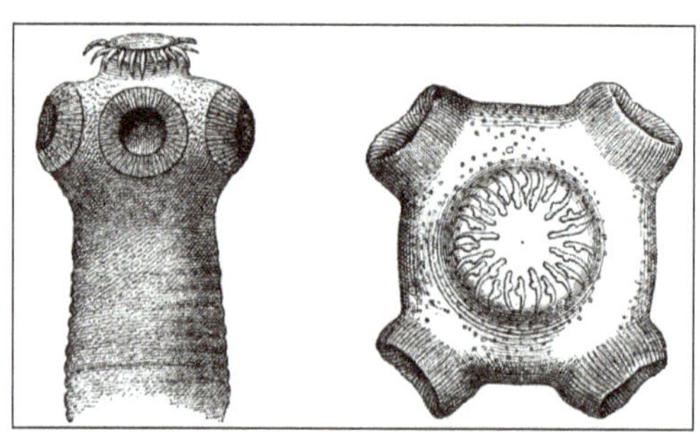

Head of the pork tapeworm, Taenia solium., with attachment structures to attach to the wall of the small intestine.

Malaria, caused by the apicomplexan parasite Plasmodium falciparum, is an intracellular endoparasite. This parasite relies on a third organism, in the form of an Anopheles mosquito vector. The host blood provides an ideal rich source of glucose and amino acids to the parasite, particularly during blood stage infection where Plasmodium infects erythrocytes. To acquire essential nutrients Plasmodium has to compete with both the vertebrate and insect host and therefore must be highly efficient, regulating uptake according to nutrient availability. Plasmodium, along with many other endoparasitic parasites, have numerous channels in their parasitophorous vacuole membrane rendering it permeable to organic solutes to allow the uptake of necessary nutrients. The Plasmodium falciparum hexose transporter (PfHT) is such a transporter, which is critical for the uptake of glucose and fructose and therefore survival of the parasite. These organic molecules have to cross three membranes altogether; the plasma membrane of the erythrocyte, the parasitophorous vacuole membrane and the Plasmodium plasma membrane, facilitated by transporters such as PfHT.

Ectoparasites

Ectoparasites are ubiquitous, often highly damaging and in most cases cannot be permanently eradicated; hence, they must usually be managed at a local scale with insecticides or endectocides. However, the growth in resistance, the slow rate of development of new actives, coupled with environmental and health concerns associated with the continued use of some of the existing neurotoxic insecticides, suggest that more sophisticated approaches to their management need to be identified. These approaches need to allow ectoparasite populations to be maintained at acceptable levels, while conserving the compounds that remain available. The development of integrated approaches, in which cascades of management tactics are deployed, with parasiticides available as one component to be used in requisite circumstances, may be the most appropriate route to achieving this aim. An essential element of such an approach is the clear articulation of the purpose of intervention and rational justification of the time-point and manner in which it is attempted. However, for this to be possible, considerably better information is required about the effects of ectoparasite abundance on animal welfare and productivity, in addition to the greater availability of effective alternative control tools.

Ectoparasites, such as many of the lice or mites, which live in relatively permanent association with their hosts, are usually small and highly susceptible to factors such as desiccation and have relatively low mobility. The risks and uncertainties associated with living without their host and finding another meal are sufficiently high that for these animals excessive virulence, leading to the death or debilitation of the host, might result in their own death and failure to reproduce. Hence, these ectoparasites are generally obligate, host-specific specialists, in normal circumstances, doing minimal damage and, in some cases, existing almost as commensals, e.g. species of Demodex mites. In many cases, these arthropods may well have followed the first evolutionary route to ectoparasitism; even before becoming ectoparasites they were preadapted to living in close association with vertebrates and their survival and dispersal depended on the continued existence and health of the vertebrate with which they were associated.

In contrast, ectoparasites that are relatively mobile, so that they are able to find a new host quickly and efficiently, are relatively resistant to the adverse effects of climate so that they can survive without the host, and have a broad range of relatively abundant hosts, might be expected to evolve higher levels of virulence allowing them to maximise the resource gain from a host as quickly as possible. For these ectoparasites, the death of the host may be of little importance since they are able to survive well independently and find a new host quickly when needed. Arthropods with these characteristics, which inflict relatively high levels of damage, such as many of the blood-feeding flies and ticks, may have followed the second evolutionary route to parasitism, starting off as free-living scavengers or predators which subsequently became opportunistic feeders on vertebrates.

Of course, varying degrees of virulence between the extremes described will have evolved, depending on the precise interactions between the ectoparasite and host.

Domestication and Intensification

Regardless of the nature of the co-evolved relationship between parasite and ancestral host, in relatively recent history these associations have been subjected to a number of disruptions which, in some cases, have resulted in a substantial transformation of the established balance of the parasite–host relationship.

In particular, the selection of desired physical and behavioural traits in companion animals and in livestock, the selection for more productive domesticated breeds combined with the intensification of production and husbandry practices, has resulted in a greatly increased susceptibility to parasites. In some cases, it appears likely that current major ectoparasite problems are very largely the result of the artificial selection for productive livestock or companion animals with specific growth and conformational characteristics. For example, selection of sheep for the thicker underwool has resulted in modern sheep breeds, which are far more susceptible to a range of parasites, such as lice, mites and blowfly strike. This is exemplified by the sheep scab mite Psoroptes ovis, where phylogenetic studies suggest that in wild, more established hosts, the parasite exists largely as an ear mite with relatively low pathogenicity; whereas, the highly pathogenic body form of the disease occurs more commonly in more susceptible domesticated hosts. The dermatological problems of domesticated dogs or malocclusion in domesticated rabbits, with resultant implications for increased ectoparasite susceptibility, provide similarly striking examples.

Ectoparasite Eradication

Under some circumstances, such as on islands, local eradication of specific ectoparasite species can be achieved and can be maintained if import controls are tight and surveillance is active, as shown for a range of major diseases in New Zealand, sheep scab in Australia and warble fly in the UK. Eradication can also be achieved for exotic parasites, such as screwworm in North Africa or when co-ordinated (but expensive) continental campaigns are driven by government legislation with comprehensive extension service support, such as for screwworm in the USA.

However, for non-isolated populations, in most circumstances complete eradication is not a practical, sustainable or economically viable option. The persistent pressure of immigration results in rapid breakdown within control areas. Furthermore, without central coordination, the limited and geographically inconsistent manner in which insecticides and endectocides are used by individual farmers results in, at best, temporary local control and necessitates more frequent repeated application. Clearly for many ectoparasites, ongoing management incorporating repeated treatment, rather than eradication, is likely to be the only realistic prospect. This has profound implications for the approaches that can be used sustainably for ectoparasite management.

Until the end of the 19th century, where it was attempted, pest and parasite control depended almost entirely on relatively inefficient but low cost management practices such as pasture hygiene, pasture rotation, crutching, dagging, clipping and the use of natural plant derived insecticides, sulphur, dusts, oils, and highly toxic heavy metals. The first commercial sheep dip was introduced in 1843, when the first wettable powder containing arsenic and sulphur was marketed by William Cooper. Since then, and particularly following the development of synthetic neurotoxins in the 1940s, the relatively low cost, ease of application, speed and efficacy of insecticides rapidly propelled them towards being main weapon in the armoury applied against ectoparasites. Increasingly, however, in recent years, concern over the human health and environmental contamination associated with use of some of the older neurotoxic compounds, such as the organochlorines, organophosphates and pyrethroids, has seen their gradual withdrawal and restriction in some areas of the world. This has been counterbalanced by the development of new actives, such as the phenylpyrrazoles, chloro-nicotinyl insecticides, semicarbazones and the use of compounds such as the insect growth inhibitors and their derivatives. Nevertheless, it is well documented that the rate of development of new actives has reduced to the point where the launch of a novel product is rare. The concern surrounding this issue is heightened by the rapid development and spread of insecticide resistance. Widespread insecticide resistance now exists in most major groups of ectoparasite.

Clearly, a major imperative is now the need to conserve existing actives, and use them in such a way that they are retained for effective use when necessary. The need to reduce the rate of development of resistance and to minimise the environmental contamination that may result from repeated application of some classes of compound, suggests that more sophisticated approaches to the management of parasites and the use of parasiticides are required. An essential component of such approaches is the clear articulation of the purpose of intervention and rational justification of the time-point and manner in which it is attempted. Is control of particular ectoparasites really needed? If so what is the desired outcome and when should it best be applied to achieve this?

Importance of Ectoparasite Abundance

Arthropod parasites are usually relatively small in relation to the size of their host. Each individual consumes only relatively small amounts of food per day, does relatively little damage and, alone, causes relatively little nuisance. Although even at low densities, by definition, ectoparasites are responsible for some injury, in the vast majority of cases, the damage each does is difficult to measure and is judged to be sub-clinical. At such sub-clinical levels, their importance may stem more from their epidemiological significance; with infested individuals acting a source of further infection. Intervention against such low level populations may be considered unnecessary.

For most ectoparasites, the damage they do and, hence, their clinical significance is a direct result of their abundance. Most of the species that are important ectoparasites, have short generation times, produce large numbers of offspring and have very high potential rates of population growth; they are well adapted to exploit situations in which there is a temporary superabundance of food. As intensity of infection on a host increases, the effects of the parasitism move along the continuum from sub-clinical to clinical and the degree of damage increases. At some point along this continuum a decision point is reached at which intervention must occur.

The key question, of course, is when and how should this be judged? The intervention point depends on the expression of parasitism as a clinical condition which is largely a function of the number of parasites present, their potential rate of increase, their clinical effects of each parasite on the host and the effectiveness of the host response, particularly its immune function. The timing of intervention must also be affected by the tactics to be used and, in some cases, by their cost. Factors such as social acceptability and changing perceptions of welfare issues are also of growing significance.

Welfare and Intervention Thresholds

Whenever an animal is parasitized, it can be argued that by definition its health and welfare are compromised relative to when there is no disease. However, the full extent of any effects of low or intermediate levels of parasite abundance on the welfare of the host can be difficult to measure and in many cases are poorly understood. A better understanding of these issues is essential to allow rational approaches to intervention to be developed.

Even where the presence of low levels of ectoparasite abundance are judged to have little immediate impact on welfare, if the parasites are highly likely to predispose the host to further infestation resulting quickly in a clinical disease, such as with cutaneous myiasis, then the intervention threshold concept is likely to be inappropriate and infestation should trigger immediate intervention. If there is a high likelihood of a rapid and predictable increase in abundance leading to clinical disease, such as with psoroptic mange, intervention against low-level ectoparasite populations may also be indicated. Similarly, if repeated feeding by even low levels of ectoparasites trigger hypersensitivity, for example, with fleas in companion animals, or if ectoparasites are vectors for significant levels of disease transmission, or are zoonotic, again immediate control may be essential. The conclusion is, therefore, that for many ectoparasites, the concept of an intervention threshold below which they can safely be ignored cannot be applied and infestation requires rapid intervention. For others however, such as gasterophilosis in equids, oestrosis in sheep or pediculosis in cattle there may well be critical thresholds where an economic cost can be placed on intervention and below which the cost of damage inflicted by the parasites is lower than the cost of control. However, this needs to be carefully evaluated in the light of future studies, since as yet there is inadequate understanding of the impact of these parasites on the welfare of the host.

Approaches which deal with parasitic disease, particularly in livestock, entirely as an economic issue and ignoring the welfare implications, will increasingly be viewed as inappropriate, since welfare concerns will override pureanalysis of costs, as is already the case in many areas of the world.

Complexity

A final key issue to be considered is that a parasite and its host do not exist in isolation. Ectoparasite abundance and activity may be affected by a wide range of abiotic factors, such as climate and a wide range of other organisms (predators, pathogens and competitors), both directly on the host or in the wider environment or indirectly, mediated through the immune response of the host. Hence, attempts selectively to remove one component of a community may have a complex range of unintended adverse consequences. The presence of at least some level if infestation may stimulate acquired immunity making the host less susceptible to future infection. Paradoxically, in some circumstances, more intense vector control may actually exacerbate disease incidence, by leading to a loss of endemic stability.

The processes of control may also have unintended wider consequences, for example, excreted un-metabolized insecticides and endectocides may have environmental impacts, for example, in dung retarding its decomposition and depleting the abundance of dunginsects. Or the use of insecticides, for example, against biting flies may inadvertently select for resistance in other ectoparasite populations, such as ticks. Alternatively, the removal of compulsory national prophylactic treatment of sheep against psoroptic mange in UK has resulted in an unintended rise in the prevalence of lice.

Ectoparasite Management: New Approaches

In the development of approaches which allow effective management of ectoparasite populations, which minimise non-target effects and preserve the availability of our existing parasiticides, it is essential to develop more fully the use of integrated programmes. In such approaches, cascades of management tactics may be deployed as and when necessary, with parasiticides available as just one component, to be used in appropriate circumstances. The adoption of such approaches, however, will require substantive changes in the thinking about parasites by producers, owners and industry.

A wide range of new tools are becoming available to assist in this goal. These include molecular techniques, which are providing powerful new insights into diagnosis and re-interpretation of parasite taxonomy and epidemiology. Simulation modelling and particularly the use of satellite imagery are allowing a more comprehensive understanding of climatic constraints limiting spatial distribution. Botanical repellents, the use of resistant or resilient strains of livestock and a fuller understanding of the effects of better nutrition, may also all make important contributions. Unfortunately, the search for new vaccines for ectoparasite control remains elusive.

Biological control, off-host trapping systems and the identification and the possibility of selective treatment of susceptible individuals are three approaches.

Biological Control

The use of naturally occurring biological pathogens, such as nematodes, bacteria, fungi and viruses offer a particularly interesting approach to ectoparasite management. For example, the common bacterium, Bacillus thuringiensis has been used on sheep in small-scale field tests for the prevention of blowfly strike in Australia, lice of poultry, sheep body lice and buffalo fly. However, so far the use of B. thuringiensis has not achieved a wider field application. House-fly and stable-fly control with parasitic wasps on feedlots has been shown to be highly effective.

The use of fungal pathogens for the control of agricultural insect pests has been widely considered. However, only a small number of studies have considered their use against arthropod ectoparasites. Previous work had shown that the fungal pathogen Metarhizium anisopliae can infect ticks on livestock, giving mortality in experimental trials of 30 and 37% for Rhipicephalus appendiculatus and Amblyomma variegatum, respectively. In field experiments, M. anisopliae was shown to induce mortalities of 83% in R. appendiculatus which were naturally infesting Zebu cattle. Treatment of Ixodes scapularis with M. anisopliae induced 100% mortality in females and engorged larvae, respectively. Similar results were reported by Samish et al. who tested several isolates of M. anisopliae against Rhipicephalus sanguineus and found the most successful isolate caused 82.6 and 60% mortalities in engorged larvae and nymphs, respectively and 92–100% mortality in adults and unfed larvae and nymphs.

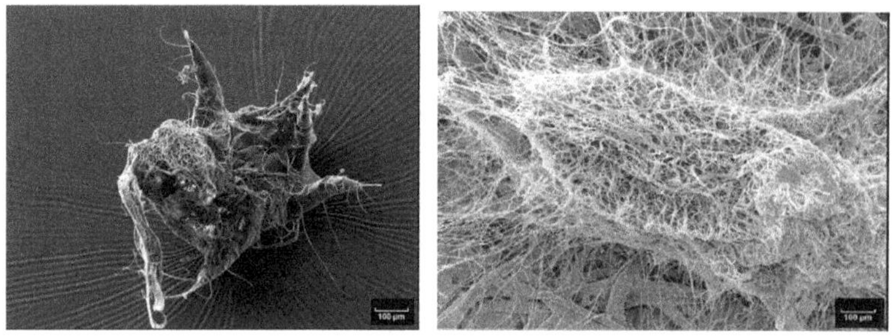

Adult female Psoroptes ovis (syn. P. cuniculi) infected by Metarhizium anisopliae. Upper—24 h after the initial observation of infection, hyphae can be observed protruding from the cuticle. Lower—5 days after the initial observation of infection, the cadaver is fully colonised by the fungal pathogen, which has started to sporulate.

In initial studies, our work sought to determine whether similar acaricidal effects could be achieved with M. anisopliae against the sheep scab mite P. ovis. For this a series of laboratory assays were used to show that the fungal pathogen is highly pathogenic to Psoroptes mites and that the number of mites which developed infections increased significantly with the increasing concentration of the conidial suspension to which they were exposed; 77% of mites developed infections when exposed to the highest concentration used (1×10^{8} conidia m l^{-1}). Controls developed no fungal infections. Mites allowed simply to walk across a surface which had been treated with a suspension of conidia, also acquired fungal infections. After contact for 24 h with a surface treated with conidia, 73% of the mites became infected.

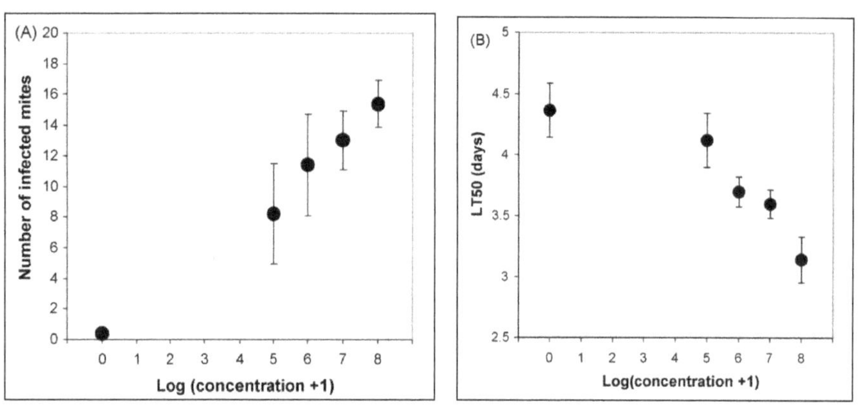

(A) The mean number (± S.E.) of adult female P. ovis (N = 20), derived from rabbit hosts, with fungal infections after exposure to a range of concentrations of the conidia from cultures of M. anisopliae. (B) The mean time taken (± S.E.) in days for adult female mites to reach 50% mortality levels (LT_{50}) after exposure to a range of concentrations of M. anisopliae.

If mites become infected and die on the host, it seems possible that cadavers will form a continuous reservoir of conidia on the host, contributing to the residual preparation applied to the sheep. The effectiveness of the material therefore might be expected to accumulate on the host rather than decay as with conventional insecticides, at least in the initial stages following infection. The source of infection also could be dispersed over the body of the host by infected, but not-yet-dead, mites and infected mites could also be transferred from sheep to sheep, further spreading the infection. To determine whether this was likely and whether dead infected mites could act as sources of infection, in experiments, live uninfected mites were placed in small chambers with infected cadavers. This showed that dead infected mites could infect naı¨ve uninfected mites; 5- day-old cadavers were the most infective and 18-day-old cadavers the least infective. However, when contact was prevented, using a mesh cage which would allow the passage of conidia but exclude direct physical contact none of the uninfected mites showed signs of mycosis-induced mortality. This showed that direct contact with the cadaver was required for infection.

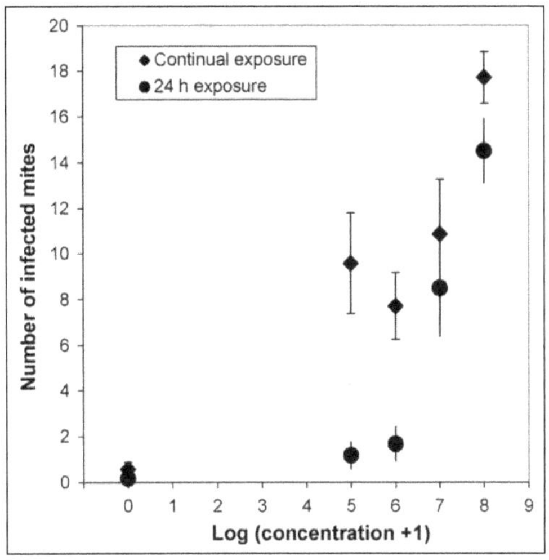

The mean number (± S.E.) of adult females of the mite Psoroptes ovis (N = 20), derived from rabbit hosts, with fungal infections after exposure to surface treated with a range of concentrations of the conidia from cultures of M. anisopliae for 24 h and continual exposure.

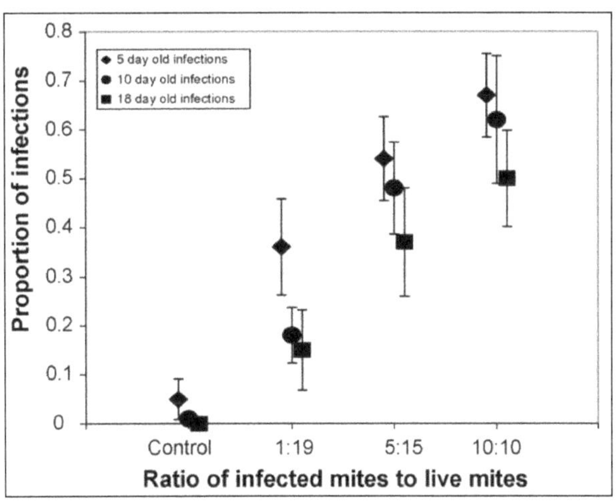

The proportion (± S.E.) of adult females of the mite Psoroptes ovis derived from rabbit hosts, with fungal infections after exposure to infected mites with different ages of infections (5, 10 and 18 days old) and in different ratios (19 live mites:1 dead infected mite, 15 live mites:5 dead infected mites, and 10 live mites:10 dead infected mites).

These laboratory results indicated that M. anisopliae is a good candidate control agent for Psoroptes mites. More recent work, has now demonstrated that in controlled applications to sheep both M. anisopliae and Bauvaria bassiana can induce high levels of mycosis in Psoroptes mites. To allow the development of an effective treatment, work now needs to address technical factors such as the development of formulations which maximise the period of residual activity and, particularly which allow good penetration of the pathogen through fleece to reach the mites on the skin surface. At present we know that the conidia will remain active in the sheep fleece for at least 3 weeks, but many of the important practical issues such as the stability of the formulations or the effects of weather and particularly exposure to UV light are unknown.

Infected adult Bovicola bovis (figure courtesy of Laura Briggs).

The effects of M. anisopliae against the common louse of cattle, Bovicola bovis has also been evaluated. Two different concentrations and formulations of conidial suspensions were used to treat lice contained in 7 cm diameter circular arenas glued to the backs of Holstein cattle, maintained in controlled climate conditions. The treatment with M. anisopliae resulted in high levels of mortality and infection with no difference between the two formulations (Tween 80 or silicone oil) in the number of infections observed. At the highest concentration (1×10^8 conconidia $m\,l^{-1}$) a mean of 73% ($\pm$ 15.57) of the lice became infected. The results suggest that the strategic seasonal use of a fungal pathogen on cattle, applied in early winter, may be of value in suppressing the winter increase in abundance, preventing the population increasing to clinically significant levels. Certainly, it appears likely at present that the development of the fungal pathogens for use on cattle is likely to be possible more quickly than for sheep, probably because it is likely to be easier develop formulations to penetrate through the hair of cattle to reach the parasites.

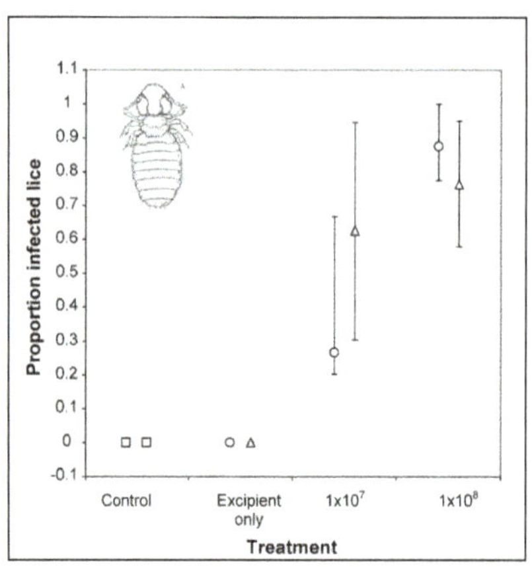

Median proportion of lice infected ($\pm$ interquartile range) after 7 days after treatment with a spray application of M. anisopliae suspended in either 0.03% Tween 80 (○) or Silicone oil (Δ) at concentrations of 1×10^7 or 1×10^8 conidia $m\,l^{-1}$ on cattle in-vivo. For controls the lice were untreated (□) or treated with excipient only. Each group contained 20 lice and treatments were replicated 5 times for the 1×10^8 and 4 times for the 1×10^7 conidia $m\,l^{-1}$ suspensions.

In terms of its environmental acceptability, M. anisopliae is classified as a low risk pathogen, which is unlikely to cause human or animal disease. Its safety to vertebrates therefore makes it a good candidate for ectoparasite treatment. However, it is notable that in the studies described, while the pathogen often gave high levels of mortality, it was never complete, regardless of the concentration or delivery method used. This is a common feature of biological control agents which, for most pathogens, can be used to suppress but not eliminate the parasite. Hence, to be used most effectively and appropriately biological control agents must be incorporated as one component of a disease management system. It should also be noted that the fact that some mites do not acquire infections suggests that there is the potential for the development of resistance, if shown to be the case, this would need to be managed appropriately.

Off-host Control

Considerable advantage may accrue if the amount of time spent handling animals for parasite control can be minimised and if approaches can be found which allow parasites such as ticks or flies to be attracted and killed in appropriate numbers during the stages in which they are present off the host.

The control of populations of pest insect species using non-return traps and targets (screens), usually accompanied by semiochemical baits, has been considered widely. As a component of the eradication of the screwworm fly, C. hominivorax, from North America, trapping using baits and insecticide was shown to reduce fly populations by about 65–85%. A noninsecticidal trap for horn fly, Haematobia irritans, developed in Australia and tested in Florida and Alabama, found that farms using traps had a 85–97% reduction in fly abundance compared to control farms with no traps, and that trapping removed the need to use insecticides to control this pest on milking dairy cattle. Nevertheless, the large numbers of adult females that need to be attracted and killed to achieve effective population management is often not achievable with the visual and olfactory baits available. As a result, for many pest insect populations, traps or targets are usually only used as monitoring tools. One notable exception to this is tsetse fly, Glossina spp. where the high level of control that can be achieved is due in large part to their very low rate of reproduction and the availability of highly effective baits and traps.

Recent studies, however, have suggested that odourbaited catching devices may be able to make an effective contribution to the control of cutaneous myiasis of sheep. Bins containing insecticide-impregnated carrion baits, modified West Australian blowfly traps, carrion-baited targets impregnated with the insect growth regulator triflumuron and traps with synthetic lures, have all been used both to suppress fly populations and to reduce strike incidence.

In Australia, a non-return insecticide-free trap for catching Lucilia cuprina has been developed and is now commercially available. It is supplied with a synthetic chemical lure formulated to attract for L. cuprina, composed of butyric acid, 2-mercaptoethanol, indole and 20% sodium sulphide solution, known as LuciLure. The ability of this trap and bait system to suppress fly populations and to reduce strike incidence has been investigated in the southern hemisphere with variable results, although reductions in strike incidence of up to 46% have been reported.

The use of fly-traps for the control of sheep blowfly strike by Lucilia sericata has also been considered extensively in the UK. Recently, trials were undertaken on 12 commercial sheep farms in southwest England. Two flocks acted as controls, with no prophylactic blowfly strike control. Four flocks of lambs and three flocks of ewes were protected only by blowfly traps. These simple devices contained sticky strips to capture flies which enter the trap and rehydrated freeze-dried liver as bait. Seven of the flocks of ewes and six flocks of lambs were protected by blowfly traps, but the farmers also chose to apply parasiticides to the animals, as they considered appropriate. Strike incidence was monitored over summer on all these farms. The results showed that the highest incidence of strike was seen in the two control flocks where 10.9 and 5.8% of the ewes and 10.1 and 9.2% of the lambs were struck. Strike incidence in the flocks that used trapping only and flocks that used trapping plus a chemical preventive was on average five times lower than in the control flocks, and the percentages struck on the sites in the trappingonly and trapping-plus-insecticide groups were not significantly different from each other. The results demonstrate that traps can make an effective contribution to blowfly strike control on individual farms. A modified version of this trapping system has now been made available commercially in the UK, sold as Agrilure®. Notably the level of protective effect achieved by the UK trap was similar to that reported by the traps used in Australia, and clearly as with biological agents, traps do not give complete protection. This further emphasises the point that these tools must be viewed as one component an integrated management programme, and unlikely to be used as stand-alone systems.

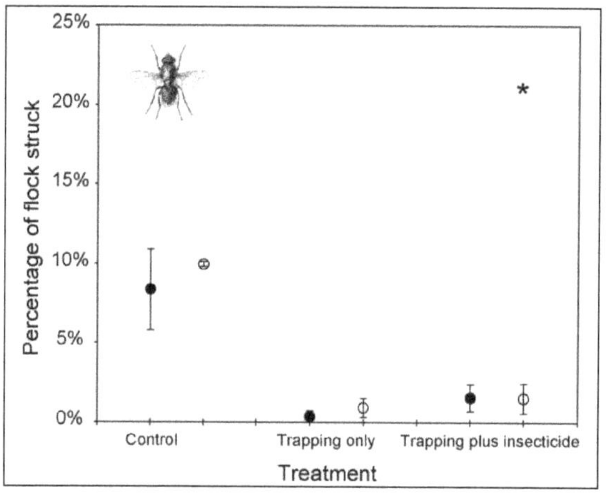

The mean (± S.E.) percentage of ewes (solid circles) or lambs (open circles) struck in three treatment groups (no control, trapping only and trapping plus insecticide) in a regional trial involving 12 farms in southwest England. One farm was considered to be statistical outlier (*) and excluded from the analysis in the lamb treatment group; this farmer claimed to have treated all sheep with cyromazine, but still had more that 20% of all lambs struck during the period when they should have been protected.

Finally, trapping systems and biological agents can be combined. Trials have been undertaken to test whether it was possible to use the fungal pathogen M. anisopliae against L. sericata. Adult female L. sericata were exposed to suspensions of fungal spores on treated cloth surfaces. The duration of tarsal contact had a significant effect on levels of infection. Furthermore, mean infection levels of 30% were observed following exposure of free-flying adults to a 5 cm × 15 cm cloth surface which had been treated with conidia and suspended from the roof of a cage (30 cm cube). Hence, it is entirely possible that fungal pathogen could be formulated for use in the field on targets or within a trap.

Targeting Susceptible Individuals

It is a well-recognised feature of many host–parasite systems that the parasite population is typically distributed within the host population in an overdispersed or clumped manner. As a result, the majority of hosts carry relatively few parasites while a small number of heavily infected hosts carry high parasite burdens, comprising the majority of the parasite population. Such distributions have been observed for virtually all parasite taxa and this aggregation is most often described by a negative binomial distribution.

The presence of highly aggregated parasite burdens may result in increased density-dependent effects on parasite size, fecundity and survival and may increase the probability of parasite-induced host mortality. Hence, reports of mean average levels of intensity of infection are often of little bearing and clinical and sub-clinical levels of parasitism can occur simultaneously within the same group of animals. Furthermore, individual hosts may vary in their clinical response to infestation. Some resilient individuals may tolerate high burdens with few signs, while others may high highly susceptible. Differences in susceptibility may be associated with age-class. This variation may reflect differences in immune response or behavioural tolerance. Variation in parasite load and host susceptibility may be causally linked (but this is not always necessarily the case). Parasite aggregation also has implications for diagnosis; the number of animals that should be sampled to estimate a critical level of infection will be highly dependent on the level of aggregation.

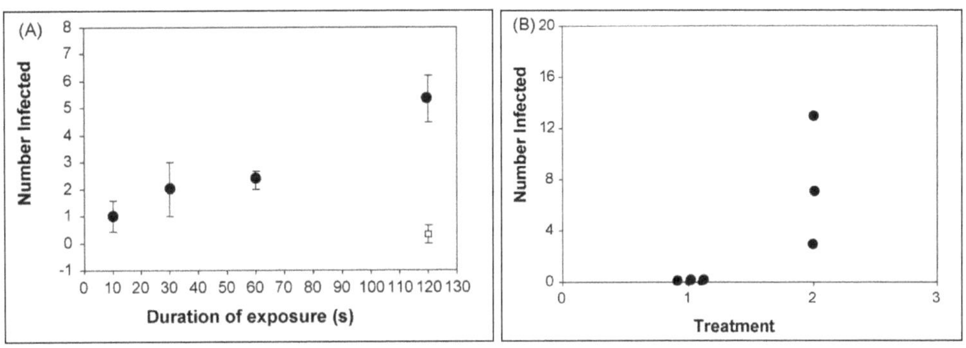

(A) Mean number (± S.E.) of adult female Lucilia sericata infected within 10 days following tarsal contact with cloth treated with 100 ml of a suspension of 1×10^8 conidia $m l^{-1}$ of M. anisopliae in silicone oil for durations of 10, 30, 60 or 120 s (solid symbols) or cloth soaked in silicone oil alone prior to tarsal contact for 120 s (open square). Each group contained 10 adult flies. (B) Number of adult female Lucilia sericata infected within 10 days, from groups of 25 flies exposed over 10 days to cloth targets. Treatment 1: targets treated with 2 ml silicone oil only. Treatment 2: targets treated with a suspension of 1×10^8 conidia $m l^{-1}$ of M. anisopliae in silicone oil.

Knowledge of the degree and causes of parasite aggregation may be of considerable value in allowing more efficient and cost-effective control strategies to be implemented (Birkett et al., 2004). At high degrees of parasite aggregation, selective control of only a few highly infested individuals will be needed to achieve effective suppression of the parasite population. It has been suggested that, at parasite aggregations described by a negative binomial distribution, with $k = 0.05$, a 50% decrease in the parasite population can be achieved by treating only 8% of the host population. Conversely, if parasites are highly aggregated, some hosts may act as a refuge for a large proportion of the parasite population. Hence, failure to treat these heavily infested animals will mean that treatment may have less effect on the parasite population than might be expected, were parasites distributed between hosts at random.

Clearly in some circumstances, the ability to identify and selectively treat more susceptible and heavily infested individuals, will be of considerable value since it will result in lower numbers of treatments, with consequent financial gains to the farmer and the potential for reduced resistance development. This strategy may be particularly valuable, where low parasite burdens are acceptable. In sheep for example, blowfly strike by L. sericata is often highly aggregated within specific age-classes at particular time of year and on more susceptible individuals. Once struck, infested animals become highly attractive to other ovipositing female blowflies and thus are more likely to receive multiple ovipositions. Using simulation models, we have shown that if the factors that predispose sheep to strike can be identified and these highly susceptible individuals treated in such a way that the larvae are killed without reducing the susceptibility of the host to oviposition, the increased aggregation of strikes would enable control efforts to be focused on these highly infested individuals.

For this approach to be used, techniques would need to be developed that can allow these individuals to be identified so that treatment can be applied specifically. However, one note of warning: our simulation analyses suggest that if these more susceptible individuals are rendered less susceptible, flies are likely simply to shift their oviposition to the next most susceptible individuals within a flock. Hence, it is important that the approach should treat the most susceptible individuals in a way that makes them lethal but not repellent to the parasites, so flies continue to be attracted, effectively turning them into living traps. Hence, the treatment should not disrupt the aggregated nature of the disease.

One valuable extension of this approach is the oncept of the selective application of insecticide to particular sites on the host, either focussing on sites where hosts such as cattle are more likely to be bitten, for example, by tsetse or where initial parasite colonisation is likely to be initiated, for example, through the use of insecticidal footbaths to prevent infestation by A. variegatum ticks. Both of these approaches allow for the more focussed use of insecticide.

Parasitology

Parasitology is the study of parasites, organisms that live, grow, and feed on or in other organisms. The prevention of parasite-infested consumption of raw (or undercooked) meat, fish, seafood, vegetables, and dairy products, as well as contaminated water, is a matter of public health. The United States Food and Drug Administration (FDA), the Centers for Disease Control (CDC), and several other local and state sanitary agencies are responsible for regulatory food safety measures and regular inspections of food and water quality to prevent the outbreak of epidemics caused by parasites and other pathogens.

When an epidemic outbreak occurs in a city or when several cases of food-related poisoning suddenly happen in an area, forensic pathologists or forensic parasitologists help epidemiologists to identify the source of the problem. For example, in 1980, 32 patients, including four physicians, reported to hospitals in Los Angeles within a short period of time complaining of abdominal distention, diarrhea, intermittent abdominal cramps, and flatulence. They were diagnosed as having been infested by a flatworm, Diphyllobothrium spp., a common parasite in freshwater and sea fish. All patients recalled that they had eaten sushi, a raw fish dish, ten days prior to the onset of symptoms. Alerted by hospitals, the CDC tracked the illness back to sushi made of salmon contaminated with the flatworm.

Another field where parasitology is also important is legal medicine , as some parasitic pathogens (disease-causing organisms) are transmitted through sexual contact and may constitute evidence of crime, especially in cases of child molestation.

Although some pathogenic (disease-causing) bacteria such as Chlamydia and Ricketsia can be thought of as obligate intracellular parasites (i.e., they can only be replicated inside living cells using the host cell's metabolic machinery) the strict definition of parasites refers to protozoa and helminthes or worms, also known as Metazoa. Pathogenic protozoa are unicellular (e.g., single-celled) organisms divided into four groups: Sarcodina (amoebas), Sporozoa (sporozoans), Mastigophora (flagellates), and Ciliata (ciliates). Metazoa or worms classified are divided in two groups, Platyhelminthes or flat worms, such as Trematoda (flukes) and Cestoda (tapeworms), and Nemathelminthes or roundworms. The most commonly occurring parasites in humans can be also grouped according to the areas of the body they infest, such as: 1) the intestinal tract (Giardia lamblia, Entamoeba histolytica, and Cryptosporidium); 2) urogenital tract (flagellate Trichomonas vaginalis); 3) blood and tissues (flagellates Leishmania and Trypanosoma, protozoans Toxoplasma and Plasmodium).

Giardiasis, or infestation by Giardia lamblia, occurs in two forms: Giardia trophozoites (active Giardia) and cysts (latent, non-mobile Giardia). Water and food contaminated with fecal residues are the main means of transmission, with the cysts developing into Giardia trophozoites in the duodenum (upper part of the stomach). Giardia attaches to the duodenal mucosa where it competes for protein and fat nutrients, causing inflammation, flatulence, foul-smelly diarrhea, intestinal cramps, nausea, anorexia, and associated protein and fatty acid deficiency. Although 50% of the hosts do not present with symptoms, giardiasis is very common among children in daycare centers, and people who camp, hike, or drink unfiltered water directly from streams, with symptoms appearing especially in those with certain immune deficiencies. Giardiasis is an endemic infestation in the United States, affecting about 5% of the population.

Entamoeba histolytica have two life-cycle phases: trophozoites or mobile amoeba and cyst (non mobile) phases. They cause intestinal cramps, dysentery, and liver lesions, being transmitted by ingestion of cysts present in water or uncooked food, as well as through fecal-oral contact in sexual intercourse. Once inside the body, the cysts mature to the trophozoites phase, the active ameba. By causing necrosis (cell and tissue death and decay) of the intestinal epithelium, amebas invade the submucosa layers of the colonic tract and reach circulation, being transported to the liver where they cause systemic hepatic disease and liver abscesses. Approximately 2% of the American population suffers from amebiasis. Other types of amebiasis are rare, such as those caused by Acanthamoeba ssp. and Naegleria fowleri, which are pathogenic free-living amebas transmitted by water inhalation (while swimming) and by air. They can multiply in the tissues of the brain and spinal fluid, causing nerve damage and death if untreated. Naegleria causes primary amoebic meningoencephalitis (PAM) and Acanthamoeba leads to granulomatous amoebic encephalitis (GAE). If untreated, PAM can kill within a week of the onset of symptoms. GAE occurs in patients with immunodeficiencies and leads to death within several weeks to a year after the onset of disease. Both diseases cause eye infections that can lead to blindness. Between 1985 and 1986, 22 cases of amoeba-related ocular lesions were reported to the Centers for Disease Control. Investigators found out that the majority of the cases were associated with poor disinfection of contact lenses and homemade saline solutions.

Cryptosporidium is another pathogen that induces diarrhea, which is more severe in small children, senior patients, and those with immunodeficiencies such as HIV. Transmission is generally under the form of oocysts present in water and may cause collective outbreaks of watery diarrhea with risk of severe dehydration, particularly to those belonging to the more vulnerable groups. Water filtration is the most effective way of preventing both giardiasis and Cryptosporidium-related diarrhea because these two parasites are resistant to water chlorination.

Almost two billion people live in parts of the world where malaria is an endemic (naturally occurring in the environment) disease. Malaria is a parasitic disease caused by four different species of the Plasmodium parasite, and is transmitted by the bite of infected mosquitoes. The worldwide use of pesticides containing DDT greatly reduced the incidence of malaria, but since DDT was found to contain possibly carcinogenic (cancer-causing) chemicals in the late 1960s, its use has declined greatly, and in turn, the incidence of malaria has increased sharply around the world. As of 2005, malaria is estimated to have killed more than 300–500 million people over the centuries and still kills an estimated 2.5 million people per year (including 1 million children) in Africa and the world's tropical areas. Many countries in these regions are returning to the use of DDT to control the mosquitoes carrying the parasite that causes malaria.

Trichomonas vaginalis is a sexually transmitted parasite that exists only as trophozoites, causing genital itching and smelly-greenish vaginal secretions as well as urethritis (a burning sensation when urinating). In men, the only symptom is urethritis, although the parasite is transmitted in the prostatic secretions (secretions of the prostate gland). The use of condoms prevents infection. When found in a child, this and other sexually transmitted diseases may suggest a case of child molestation. Some rare cases of trichomoniasis appear to be associated with contact with wet toilet seats.

Toxoplasma gondii, a blood parasite, may be transmitted through the contact with infected feces of cats and other mammals, or by consumption of raw or undercooked meat or contaminated water, causing toxoplasmosis. It can be also transmitted from mother to the fetus, in what is known as congenital toxoplasmosis. Congenital infection favors miscarriage, neonatal mental retardation, or chorioretinitis (inflammation of the choroids portion of the eye), which leads to blindness during childhood. In immunodepressed adults, toxoplasmosis may cause encephalitis, although most of the infected population remains asymptomatic, due to the action of the immune system . However, T. gondii passes from the intestinal tract to other tissues of the body, such as brain, liver, lungs, and eyes, where it remains as cysts for years. As long as the infected individual's immune system is healthy, antibodies and the immune cells will keep the infection at bay, preventing disease progression.

Diarrheal parasites and other pathogens account for 4% of deaths worldwide each year. Periodical tests for these and other parasitic infestations are a valuable preventive measure that can avert serious and unnecessary diseases and even death.

Complex Lifecycle Parasites

Many parasitic infections, including those of humans, are caused by complex lifecycle parasites (CLPs): parasites that sequentially infect different hosts over the course of their lifecycle. CLPs come from a wide range of taxonomic groups—from single-celled bacteria to multicellular flatworms—yet share many common features in their life histories. Theory tells us when CLPs should be favoured by selection, but more empirical studies are required in order to quantify the costs and benefits of having a complex lifecycle, especially in parasites that facultatively vary their lifecycle complexity.

Parasites vary in the complexity of their lifecycles. Some complete their lifecycle in a single host, while others require multiple successive hosts; some parasitic helminths must sequentially infect four obligate host species to complete their lifecycle. CLPs are of substantial applied importance because they cause severe diseases of humans and livestock, including malaria, toxoplasmosis, river blindness, Lyme disease, schistosomiasis and sleeping sickness. However, despite abundant and the clear relevance to human health, we lack empirical studies of the evolutionary causes and consequences of parasite lifecycle complexity.

At first glance, the evolutionary transition from simple to complex lifecycle seems unlikely, as it requires the parasite to rely on multiple host species, usually in a particular sequence, in order to complete its lifecycle. Nevertheless, CLPs abound in nature and have evolved in multiple disparate taxa, including bacteria, flatworms, roundworms, acanthocephalan worms, Arthropods and two different lineages of protists: the Apicomplexa and Kinetoplastida.

Evolving a Complex Life Cycle

Parker et al. suggested two mechanisms that may have led helminths to divide their lifecycles across hosts. The first of these is upward incorporation (UI) and requires the presence of predators that routinely eat (and sometimes actively choose) parasite-infected prey. If parasites can adapt to survive and reproduce in the predators of their hosts, they may acquire the predator as a second host. The second mechanism is downward incorporation (DI). DI evolves in two stages: first, a directly transmitted parasite evolves the capability to survive independently of its host; and second, the parasite evolves to infect a second host species that routinely ingests these parasite transmission stages, thus acquiring a second host. UI and DI can both, for different reasons, lead to increased parasite fitness: UI permits a longer parasite lifespan, greater body size and increased fecundity, whereas DI leads to reduced mortality of parasite propagules and therefore increased probability of transmission.

There are, however, constraints to evolving complexity. Choisy et al. demonstrated that selection should only favour an increase in lifecycle complexity when intermediate hosts are more abundant than the definitive host, parasite survival in the intermediate host is high and transmission from the intermediate host to the definitive host is common (that is, when lifecycle complexity increases the efficiency of parasite transmission). Increased lifecycle complexity should evolve only if the cost of being able to infect multiple hosts is low. Gandon's work supports these findings: after designing models with a malaria system in mind, he found that indirect vector-borne transmission was favoured when the supply of vectors was high compared with the supply of the definitive host.

The benefits of a complex lifecycle must be realised rapidly for it to be favoured by selection. Morand et al. tested for these benefits by comparing the fitness of two closely related parasitic cestodes that differed in the number of intermediate hosts they infect. Bothriocephalus barbatus infects a marine copepod as its intermediate host and a flatfish (Scopthalmus rhombus) as its definitive host. Bothriocephalus gregarious infects the copepod as a first intermediate host, gobies (Pomatoschistus sp.) as second intermediate paratenic (non-essential) hosts and a different flatfish, (Psetta maxima) as its definitive host. Although not strictly necessary for the parasite to complete its lifecycle, gobies likely benefit B. gregarious infectious stages by increasing overall transmission: the total prevalence of B. gregarious is higher than that of B. barbatus. However, although comparing two closely related parasite species with different levels of complexity is useful, one cannot isolate the effect of differing lifecycle complexity from other unknown phenotypic influences on parasite transmission potential; B. barbatus and B. gregarious are different in many respects in addition to the complexity of their lifecycles, for example, the identity of their definitive hosts. How can this problem be overcome? Parasites that facultatively alter the number of hosts they infect provide an excellent opportunity to test the immediate benefits of complexity, as one can easily examine relationships between lifecycle complexity, parasite transmission and parasite fitness in a single parasite species. The trematode, Coitocaecum parvum is an example of such a parasite.

Coitocaecum parvum can adopt a three-host (snail–amphipod–fish) or a two-host (snail–amphipod) lifecycle. Once in the haemocoel of the amphipod host (Paracalliope fluviatilis), C. parvum can either await ingestion by the fish Gobiomorphus cotidianus where it sexually reproduces in its gut (three-host lifecycle), or continue to grow and then reproduce by selfing in the amphipod. Coitocaecum parvum only switches to a three-host lifecycle in the presence of the third (fish) host. Although the three-host lifecycle allows cross-fertilisation and is therefore probably optimal (outbreeding provides immediate fitness benefits), the two-host strategy acts as a form of reproductive insurance, allowing the parasite to complete its lifecycle in the absence of the third host and thus avoid death.

Trials of Transmission

CLPs face the challenge of ensuring transmission between multiple host species to complete their lifecycle, and failure at any one of these transmission events could result in the death of the parasite. Parasites that can facultatively vary the number of hosts they infect (for example, C. parvum) raise many interesting questions. When such parasites are restricted to a simpler lifecycle for multiple generations, does the fitness loss associated with inbreeding accumulate? Does sustained selection on a two-host lifecycle lead to evolutionary changes in virulence (on any of the hosts) or between-host transmission?

There are abundant examples where transmission constraints have selected on CLPs to revert to a simpler (and easier to complete) lifecycle. Various trematode species have given up their definitive host evolved to use a second intermediate host as a definitive host or have foregone paratenic hosts. Poulin and Cribb note 32 trematode families have reduced lifecycles and that lifecycle truncation has occurred on at least 20 occasions in the evolutionary history of the Trematoda. Life cycle truncation is, however, often accompanied with other major life history changes—the apicomplexan Toxoplasma gondii provides a prime example.

Toxoplasma can establish infections in almost any endothermic vertebrate as an intermediate host and undergo asexual replication, but for the parasite to reproduce sexually, it must infect a felid. However, some T. gondii lineages have reverted to a simple lifecycle without a felid definitive host, and have forgone sexual reproduction as a consequence. It remains to be determined whether this switch was initially favoured by selection because of limited opportunities for transmission to felids. Lifecycle truncation has resulted in the massive clonal expansion of three asexual T. gondii lineages, and has been implicated in increased pathogenicity to humans. Does a loss of sex involve increased asexual replication, and if so, does this explain the increased pathogenicity of asexual strains? Long-term experimental evolution studies could effectively address these questions in controlled laboratory environments, and whole-genome sequencing studies comparing sexual and asexual T. gondii strains would provide an opportunity to identify the evolutionary changes underlying this marked lifecycle modification.

Parasite Manipulation of Host Behaviour

As an alternative to lifecycle truncation, constraints can select for CLP manipulation of an intermediate host's behaviour to increase transmission to the definitive host. However, infection-induced changes in host behaviour may not always be adaptive parasite strategies; they could be mere side effects of infection (for example, reduced motility) and could increase the likelihood that infected hosts are, for example, consumed by other predators in which the parasite cannot complete its lifecycle. The most convincing evidence that manipulation of host behaviour is adaptive to the parasite comes from demonstrations that the parasite (1) only induces a behavioural change at a specific time that favours its own fitness; and (2) this behavioural change increases the likelihood of transmission to a definitive host in which the parasite can complete its lifecycle. This has been examined in numerous studies. We provide a few key examples below.

Using a simple optimality model, Hammerschmidt et al. determined an optimal duration of infection for the tapeworm Schistocephalus solidus in its intermediate host (the copepod, Macrocyclops albidus). They then tested their model experimentally by altering the duration of infection in the copepod, and found immature S. solidus actively suppressed copepod activity, reducing the likelihood of premature fish predation (and death of the immature parasites); however, when S. solidus reached the next developmental stage and required transmission to the second intermediate (fish) host, they increased copepod activity and the likelihood of predation. Similarly, Dianne et al. demonstrated that the Amphipod, Gammarus pulex, exhibited stronger anti-predator behaviour (that is, was more likely to hide in predator refuges) when infected with immature stages of the acanthocephalan Pomphorhynchus laevis, than when uninfected. Once P. laevis had reached maturity within its G. pulex host, this anti-predator behaviour decreased and infected G. pulex were significantly more likely to experience predation by the definitive host, the brown trout, Salmo trutta. There is also compelling evidence that T. gondii manipulates the behaviour of its intermediate rat hosts to make them sexually attracted to the odour of cats, the definitive host.

Life History Trade-offs Faced by CLPs

Like most organisms, parasites often face a trade-off between growth and reproduction because environments that are optimal for growth may be suboptimal for reproduction and vice versa. Parasite growth (or asexual replication) is often lower when within-host competition is high (for both SLPs and CLPs). CLP coinfections (multi-genotype infections) in intermediate hosts occur in many but not all helminth taxa, and when they do occur (either naturally or experimentally), competition leads to clear parasite fitness reductions. However, sexually reproducing parasites often benefit from being in genotypically diverse infections, because they can acquire genetically dissimilar mates and thus reduce the incidence of inbreeding and the associated fitness costs. Even for CLPs that can self-fertilise, cross-fertilisation is often favourable and may even be required on some occasions in order for the parasite to maintain offspring viability and prevent a decline in overall fecundity. In definitive hosts, the benefits of outbreeding may outweigh the costs associated with within-host competition. Below, we outline a hypothesis that suggests CLPs may, in the short-term, be less constrained by the growth–reproduction trade-off than their SLP counterparts.

Short-term Escape from the Growth–reproduction Trade-off

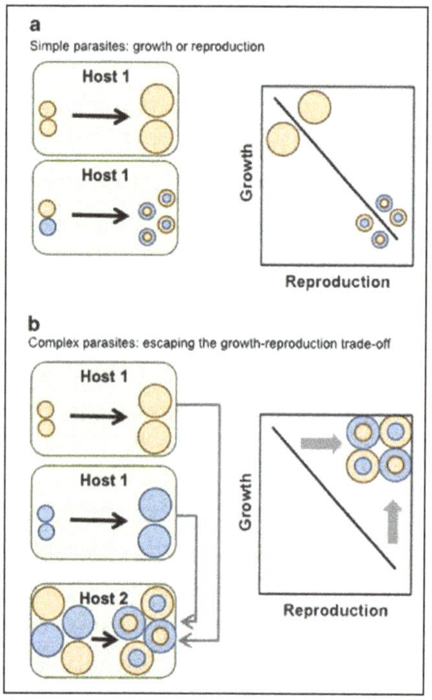

Growth–reproduction trade-off for SLPs and CLPs. Solid blue or solid orange circles represent two different clonal genotypes; circles that are both blue and orange represent the sexual progeny resulting from mating events between the blue and orange genotypes. (a) Within-host competition subjects simple parasites to the growth–reproduction trade-off: asexual reproduction leads to a few large parasites, whereas sexual reproduction leads to more small parasites. (b) CLPs can escape this trade-off: both growth and reproduction can, to a greater extent, evolve independently.

Here, we argue that a complex lifecycle may be favoured by selection because it can reduce or even eliminate the growth–reproduction trade-off (at least in the short term) by allowing the parasite to separate growth and reproduction according to host type. As figure shows, a CLP can undergo growth in intermediate hosts alongside few (if any) other genotypes, and then reproduce in definitive hosts, where there may be many parasite genotypes. Indeed, both the intensity (measured as parasite burden) and genetic diversity of infections tend to be higher in definitive hosts compared with intermediate hosts, because of concentration of parasites up the food chain. For example, in the trematode Diplostomum pseudopathaceum, a small number of parasite genotypes (usually just one) establish infections in the first intermediate (snail) host, thus optimising growth by minimising parasite competition. Immature trematodes then go on to infect the second intermediate host (fish) along with other (genotypically different) trematodes. Finally, the definitive host (a bird) consumes numerous fish and thus collects many parasite genotypes, providing an arena for the many mature trematode genotypes to mix and mate.

Contrasting selection pressures from each host type could thus allow parasite life-history traits related to growth and reproduction to evolve and adapt independently. Indeed, some cestode genera exhibit complete separation of growth and reproduction: Schistocephalus, Ligula and Digramma sp. only feed and grow in their intermediate hosts, and only reproduce in the definitive host; similarly, Plasmodium, Theileria and Toxoplasma sp. undergo asexual replication (equivalent to growth) in intermediate hosts, reproduce sexually in only their definitive hosts). This has important implications for the virulence experienced by various hosts, because virulence often increases with either parasite size or density.

The evolution of a complex lifecycle causes shifts in selection pressures on different life history stages and we acknowledge that the complete separation of growth and reproduction according to host type is not a universal feature among CLPs. For example, mortality rates are often lower in definitive hosts than in intermediate hosts, resulting in selection for growth in the definitive host before reproduction occurs. So parasite schedules of investment in growth and reproduction (and resultant changes in virulence) can be both a cause and a consequence of an increase in lifecycle complexity.

CLPs and Host–parasite Coevolution

CLPs often infect very different host types over the course of their lifecycle, for example, invertebrates and vertebrates. Moreover, many CLPs have at least one highly mobile host that promotes gene flow among parasite populations. Such gene flow can potentially limit the ability of parasites to adapt to hosts and thus prevent specialisation on either or all host types. These breaks to adaptation may lead one to predict that parasites become less specialised as their lifecycles increase in complexity. This is not the case.

There are many examples of extreme host specialisation among CLPs: Plasmodium species specialise on different species of bird, mammal and reptile, and show little evidence for host switching. Plasmodium falciparum even exhibits genetic specificity with its Anopheles gambiae (mosquito) hosts, where infection success depends on the exact combination of P. falciparum strain and mosquito genotype. Schistosoma mansoni also shows evidence of genetic specificity with its snail intermediate host, Biomphalaria glabrata, and Microphallus trematodes exhibit adaptation to locally common genotypes of their snail intermediate hosts, Potamopygus antipodarum.

Patterns of Local Adaptation in CLPs

How can a parasite have a complex lifecycle that frequently entails high between-population dispersal, and yet adapt to infect particular local genotypes of intermediate host, definitive host or both? Theory predicts parasites can show local adaptation (LA) to at least one host species (that is, achieve greater infection success of sympatric hosts than allopatric hosts), provided parasite migration rate exceeds host migration rate in a metapopulation. Put another way, the ability of parasites to adapt to their hosts (and vice versa) is contingent on their evolutionary potential, which is dependent on the supply of favourable alleles to the population.

The evolutionary consequences of definitive host migration for a CLP metapopulation. Migration rates often vary among the different hosts of a CLP. The migration rate of a CLP is expected to be most similar to that of its most mobile host—in this case the definitive host. Immigration of mobile hosts into a population can supply novel parasite genotypes, promoting parasite LA to the less-mobile host.

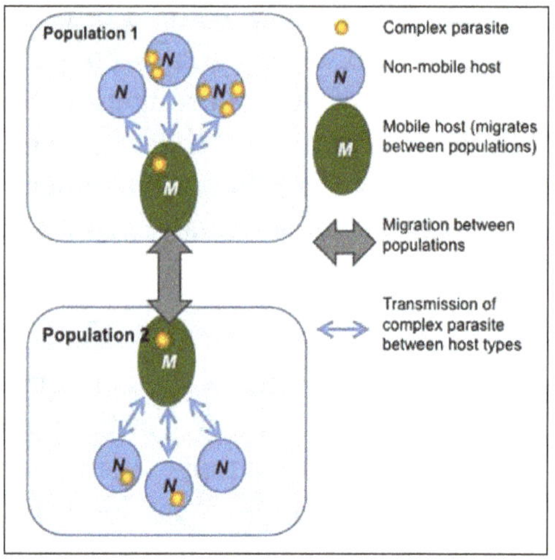

A recent meta-analysis found CLPs were no more likely to exhibit LA than SLPs. However, as discussed earlier, theory suggests that the presence of LA will crucially depend on the migration rate of the focal host species relative to the migration rate of the other host(s). This theory gives rise to the following hypotheses: (1) more mobile hosts will promote CLP dispersal, and thus increase the supply of novel genotypes for infection of less mobile hosts; (2) CLPs will exhibit LA to their least mobile host species; and (3) CLPs will exhibit either no LA or local maladaptation to their most mobile host.

Empirical studies provide considerable support for these hypotheses. Keeney et al. found that the trematodes Maritrema novozealandensis and Philopthalmus sp. exhibited higher within-population genetic diversity and greater gene flow than their low mobility marine snail host, Zeacumantus subcarinatus. Dybdahl and Lively found Microphallus trematodes had much lower between-population genetic structure than their intermediate P. antipodarum (snail) hosts. Wild Schistostoma mansoni trematodes also had population genetic signatures of dispersal similar to those of their definitive rat hosts, and S. mansoni that infect snails, salmon and fish-eating birds have higher dispersal and less population structure than those which infect snails and salmon only. These four studies suggest the more mobile definitive hosts maintain genetic variation in parasite populations, supporting hypothesis. Microphallus also exhibits strong LA to sympatric snail populations, and S. mattheei is locally adapted to its less mobile snail host Bulinus globosus, consistent with hypothesis. Conversely, Prugnolle et al. found no evidence for S. mansoni LA to the snail (Biomphalaria glabrata) in a population where S. mansoni and B. glabrata were found to have similar migration rates. Diplostomum pseudopathaceum trematodes demonstrated local maladaptation to their more mobile fish hosts (Gasterosteus aculeatus) and a haemogregarine parasite (phylum Sporozoa) showed local maladaptation to populations of its most mobile host, the lizard Gallotia galloti. These three studies support hypothesis. It is, however, important to note that many studies of LA in CLP focus on trematodes. Further study of other CLP taxa is required to effectively test whether CLPs are generally more likely to exhibit LA to their less mobile hosts.

For coevolutionary systems that rely on allelic novelty, the supply of allelic diversity to a population may accelerate evolutionary change in parasite populations by contributing alleles that evade prevailing host immune responses. Conversely, if host migration rate exceeds that of the parasite in a metapopulation, the host may have the upper hand, as higher host migration could fuel the evolution of host resistance to sympatric parasites. The potential of high-dispersal hosts to spread genotypes of CLPs and thus fuel parasite LA could pose major problems for the management of many human and livestock diseases. Limiting vector dispersal may therefore be an important component of disease management.

Mechanisms of Infection and Host Immune Evasion in CLPs

Do CLPs use the same or different machinery to infect each host type? Plasmodium berghei, P. chabaudi and P. yoelli, as well as the filarial worm Brugia malayi exhibit different patterns of gene expression when in mosquitoes than when they are in their mammal hosts. The Lyme disease bacterium Borrelia burgdorferi also expresses different genes when infecting ticks and mammals. These findings have been interpreted to indicate that the parasite uses distinct infection mechanisms to infect each host type. However, in the absence of functional knowledge, observations of differential gene expression in different hosts could merely reflect changes in parasite developmental stage, rather than host-specific infection mechanisms.

Whole-organism study has provided considerable insight into the overlap between infection mechanisms. The ability of the helminth Schistosoma mansoni to successfully infect its intermediate snail host trades-off with success at infecting its definitive mouse host. This suggests some of the genes associated with infecting snails may also influence infection success in mice and that antagonistic pleiotropic effects could maintain allelic diversity in these genes The tapeworm, Schistocephalus solidus exhibits a slightly different trade-off between its ability to overcome the innate immune system (present in both invertebrates and vertebrates) and the acquired immune system.

Studies of the molecular mechanisms of infection have further added to our understanding of CLPs. Differences in gene expression in the Lyme disease-causing bacterium, Borrelia burgdorferi are intrinsically linked with distinct mechanisms for infecting intermediate hosts (deer ticks, Ixodes sp.) and definitive (vertebrate) hosts. Borrelia expresses different suites of genes as it migrates to different tissues within the tick; upregulation of the ospA gene promotes exit from the tick gut and ospC is required for passage to the tick salivary gland and is essential for early infection in the vertebrate host. Borrelia also blocks the vertebrate complement system by producing Erp (OspE/F-like Related Protein) proteins on its cell surface and uses the saliva of its tick host as an additional shield against the vertebrate immune system (tick saliva inactivates the complement system and therefore inhibits the capacity of vertebrate neutrophils to phagocytose microbial parasites).

Plasmodium also uses different infection mechanisms for vertebrate and invertebrate hosts: both ookinetes (the mosquito-infecting life stage) and sporozoites (the mammal-infecting life stage) express genes from the TRAP-like gene family (thrombospondin-related anonymous protein), which mediate host cell recognition. The MAOP protein (membrane attack ookinete protein; a type of TRAP) is required for entry into mosquito midgut cells and the closely related SPECT1 and SPECT2 proteins (sporozoite micronemal protein essential for cell transversal) are required for successful infection of mammalian liver hepatocytes.

How could separate mechanisms for infecting different host types have evolved? The degree of overlap between infection mechanisms used to infect intermediate and definitive hosts can tell us a great deal about the evolution of lifecycle complexity in parasites, especially as infection mechanisms are likely to be linked with parasite developmental stage. Evidence from both Borrelia and Plasmodium (from Osp genes and TRAP genes, respectively) suggest that relatively recent gene duplication events have generated multiple paralogues, resulting in the evolution of two infection mechanisms from a single ancestral mechanism. If the scenario outlined in figure holds, one expects that knowledge of the infection mechanisms used for different hosts may allow characterisation of the degree of parasite specialisation and assessment of the potential for evolutionary shifts in lifecycle.

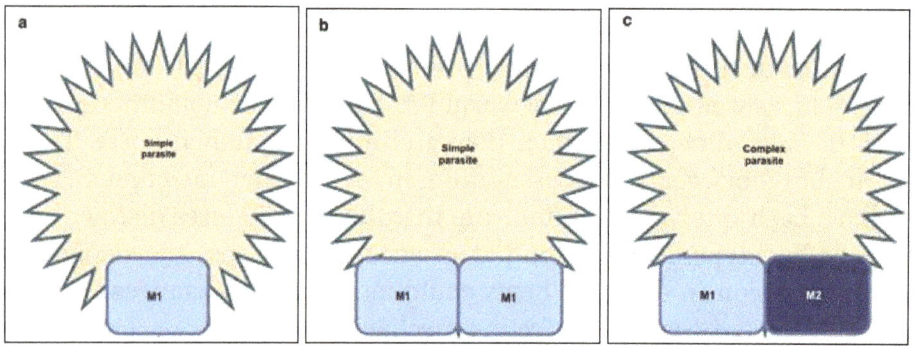

Scenario for the evolution of host-specific infection mechanisms with increasing parasite lifecycle complexity. (a) The ancestral state: a parasite with a single infection mechanism (M1) for infecting different host types. (b) Gene duplication creates paralogues of the ancestral infection mechanism: a precursor for distinct infection mechanisms. (c) Varying selective pressures from intermediate and definitive hosts lead to neofunctionalisation in one of the paralogues (M2) and the evolution distinct infection mechanisms.

Testing the gene Duplication Hypothesis and uncovering Genetic Signatures of Coevolution

We suggest gene duplication may be a common mechanism for the acquisition of novel hosts during the evolution of a complex parasite lifecycle. Examination of molecular evolution in infection gene paralogues will help address essential questions in CLP biology, for example, whether particular lifecycle complexity evolved through UI or DI. Where acquisition of a novel host has been associated with gene duplication, one paralogue will have retained its function as an infection mechanism for the ancestral host, whereas the second locus will have been subjected to strong selection for neofunctionalisation (and will show evidence of elevated adaptive substitution) as the parasite adapts to its new additional host. By looking for evidence of duplication and neofunctionalisation among genes associated with infecting different host types, one can uncover whether lifecycle complexity evolved through UI or DI. Increased lifecycle complexity likely evolved via UI if the neofunctionalised paralogue is associated with infection of the definitive host, whereas DI is more likely to have occurred if the neofunctionalised paralogue is associated with infection of the intermediate host.

Another application of sequence evolution studies would be to test the prediction that antagonistic host–parasite coevolution should be strongest between the parasite and its least mobile host. Strong host–parasite coevolution may be evident as the maintenance of divergent genotypes through negative frequency selection, or as the depletion of genetic diversity because of recurrent selective sweeps. Comparing sequence evolution in paralagous gene loci with similar infection functions could provide a strong opportunity to test the relative departures from neutrality in these two loci and hence whether the parasite has the most dynamic coevolutionary interaction with its intermediate or definitive host.

Conservation Biology of Parasites

A large proportion of living species on Earth live a parasitic way of life. Parasites have traditionally been seen as targets of eradication efforts, and they have often been overlooked in conservation efforts. In the case of parasites living in the wild – and thus harmless to humans and domesticated animals – this view is changing.

The capture, captive breeding, and reintroduction of California condors into the wild was the most expensive species conservation project in United States history. The bird was saved from extinction but its ectoparasite, the louse Colpocephalum californici, became extinct.

The captive breeding and reintroduction program to save the black-footed ferret would cause the loss of its specific parasites and demanded "equal rights for parasites!". A paper in 1992 warned that not only the loss of certain host species from the wild, but host population bottlenecks or the fragmentation of host populations would predictably lead to the extinction of host-specific parasites. The paper also noted that parasites exert selective pressures upon their host populations that increase host genetic diversity. At first, this view met with open scepticism. However, it became clear that the co-extinction of hosts and their specific parasites is likely to increase the current estimates of extinction rates significantly. A decade later, a study focusing on highly host-specific groups such as fig wasps, parasites, butterflies, and myrmecophil butterflies estimated the number of parasites put at risk by the endangered status of the host at about 6300. Other authors argued that host-specific parasite faunae have an unexpected advantage for conservation scientists. Their genealogies and population genetic patterns may help to illuminate their hosts' evolutionary and demographic history. Recently, scientists suggested that rich parasite faunae are inevitably needed for healthy ecosystem functioning and also that parasites and mutualists are the most endangered species on Earth. Even vets have started to argue about the conservational values of parasite species.

A recent study on parasites of coral reef fish suggested that extinction of a coral reef fish species would eventually result in the coextinction of at least ten species of parasites. Although this number might seem high, the study included only large parasites such as parasitic worms and crustaceans, but not microparasites such as Myxosporea and Microsporidia.

Extinct Avian Lice

The list below follows that of Mey:

- Acutifrons caracarensis parasite of the extinct Guadalupe caracara (Caracara lutosa), Guadelupe Island, Mexico;
- Longimenopon dominicanum parasite of the extinct Guadalupe storm petrel, Oceanodroma macrodactyla, Guadelupe Island, Mexico;
- Rallicola piageti parasite of the possibly extinct New Caledonian rail (Gallirallus lafresnayanus), New-Caledonia;
- Coloceras hemiphagae parasite of the extinct Norfolk Island pigeon (Hemiphaga novaeseelandiae spadicea), Norfolk Island, New-Zealand;
- Coloceras restinctus parasite of the extinct Norfolk Island pigeon (Hemiphaga novaeseelandiae spadicea), Norfolk Island, New-Zealand;
- Rallicola extinctus parasite of the extinct huia (Heteralocha acutirostris), New-Zealand;
- Philopteroides xenicus parasite of the extinct bushwren (Xenicus longipes), New-Zealand;
- Psittacobrosus bechsteini parasite of the extinct Cuban red macaw (Ara tricolor), Cuba;
- Colpocephalum californici, parasite of the California condor (Gymnogyps californianus). The host was saved by a captive breeding and repatriation program, but the parasite was lost, deliberately killed whenever it was found during the program.

Additionally, Columbicola extinctus is a parasite of the extinct passenger pigeon (Ectopistes migratorius). However, recent taxonomic studies show that it is conspecific with the lice living on band-tailed pigeon (Columba fasciata), thus it is not extinct.

Deliberate Extermination

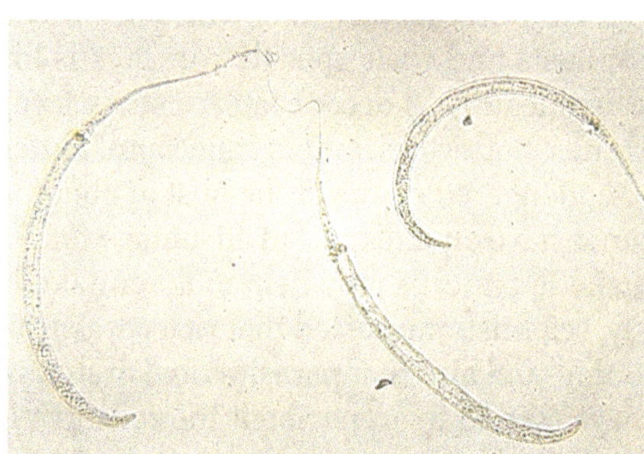

Larvae of the guinea worm: probably the next species to exterminate.

Naturally, medical and veterinary sciences have attempted to exterminate parasites and pathogens living in humans and in domesticated animals. In the case of a few highly host-specific pathogens, this means the extinction of the entire pathogen species. Throughout human history, however, only one species, the smallpox virus, has been wholly eradicated. The last reported cases of smallpox occurred 1978. However, secure samples are known to still exist in the United States and Russia, purportedly for defensive purposes such as developing new vaccines, antiviral drugs, as well as for diagnostic tests. It is not known, but is speculated that these countries may have shared their samples with some of their allies during the Cold War.

A second, potential candidate for deliberate extermination is Dracunculus medinensis, the guinea worm. Once widespread across some 20 nations of Africa and Asia, the parasite today survives only in four of the countries of Sub-Saharan Africa, with only a few hundred known cases of infection in 2011. Prevalent civil wars in the region, such as the War in Darfur, by preventing safe access to medical aid, have resulted in the survival of this species up to the present.

References

- Parasitism: biologydictionary.net, Retrieved 1 January, 2019
- "The Making of Alien's Chestburster Scene". The Guardian. 13 October 2009. Archived from the original on 30 April 2010. Retrieved 29 May, 2010
- Parratt, Steven R.; Laine, Anna-Liisa (January 2016). "The role of hyperparasitism in microbial pathogen ecology and evolution". The ISME Journal. 10 (8): 1815–1822. Doi:10.1038/ismej.2015.247. PMC 5029149. PMID 26784356
- Parasitology, medicine, divisions-diagnostics-and-procedures: encyclopedia.com, Retrieved 2 February, 2019
- Connor, Steve (2002-01-03). "How terrorism prevented smallpox being wiped off the face of the planet for ever". The Independent. Retrieved 2009-12-13

2

Branches of Parasitology

The three main branches of parasitology are medical parasitology, veterinary parasitology and paleoparasitology. These branches deal with the parasites which infect humans, parasites which infect animals and parasites from the past, respectively. This chapter has been carefully written to provide an easy understanding of the varied facets related to these branches of parasitology.

Medical Parasitology

Medical Parasitology is the science dealing with parasites that infect man, causing disease and misery in most countries of the tropics. They plague billions of people, kill millions annually, and inflict debilitating injuries such as blindness and disfiguration on additional millions. World Health Organization estimates that one person in every four harbors parasitic worms. The present work aims to clarify several aspects concerning parasites of medical importance to man. Parasite classification, general characters, biology, ecological factors that affect their transmission, the immune response of the body to invading parasites, diagnosis and control of the disease developed are highlighted.

Medical Parasitology is the branch of medical sciences dealing with organisms (parasites) which live temporarily or permanently, on or within the human body (host). Parasitology is a dynamic field because the relationships between parasites and their hosts are constantly changing. Parasitism comprises an ecological relationship between two individuals of different species where the parasite's environment is another living organism. Parasites often cause important diseases of humans and animals. For this reason, Parasitology is an active field of study in which advances in biotechnology have raised expectations for the development of new drugs, vaccines, and other control measures. However, these expectations are dampened by the inherent complexity of parasites and host-parasite relationships, the entrenchment of parasites and vectors in their environments, and the vast socioeconomic problems in the geographical areas where parasites are most prevalent. But what is the parasite?

The parasite is a living organism that lives in (endoparasite) or on (ectoparasite) another organism, termed its host. It obtains nourishment and protection while offering no benefit in return. Consequently, the host suffers from various diseases, infections, and discomforts. However, in some cases, the host may show no signs at all of infection by the parasite.

Types of Hosts

Hosts are classified according to their role in the life cycle of the parasite into:

- Definitive host (DH) that harbours the adult or sexually mature stages of the parasite (or in whom sexual reproduction occurs) e.g. man is DH for Schistosoma haematobium, while female Anopheles mosquito is DH for Plasmodium species (malaria parasites).
- Intermediate host (IH) that harbours larval or sexually immature stages of the parasite (or in whom asexual reproduction occurs) e.g. man is IH of malaria parasites. Two intermediate hosts termed 1st and 2nd IH may be needed for completion of a parasite's life cycle, e.g. Pirenella conica snail is the 1st IH, while Tilapia (Bolty) fish is the 2nd IH for Heterophyes heterophyes.
- Reservoir host (RH) harbours the same species and same stages of the parasite as man. It maintains the life cycle of the parasite in nature and is therefore, a reservoir source of infection for man. e.g. sheep are RH for Fasciola hepatica.
- Paratenic or transport host in whom the parasite does not undergo any development but remains alive and infective to another host. Paratenic hosts bridge gap between the intermediate and definitive hosts. For example, dogs and pigs may carry hookworm eggs from one place to another, but the eggs do not hatch or pass through any development in these animals.
- Vector is an arthropod that transmits parasites from one host to another, e.g. female sand fly transmits Leishmania parasites.

Host Parasite Relationship

The term refers to the relationship between the host and the parasite and the competition for supremacy that takes place between them.

Disease should not be confused with infection; a person may be infected without becoming diseased. If the host has upper hand, due to increased host resistance, it remains healthy and the parasite is either driven away or assumes a benign relationship with the host, but if the host loses the competition, a disease develops.

In biology, the relationship between two organisms is mainly in the form of symbiosis, defined as "life together", i.e., the two organisms live in an association with one another. Thus, there are at least three types of relationships based on whether the symbiont has beneficial, harmful, or no effects on the other.

Types of Symbiotic Association

- Mutualism is a relationship in which both partners benefit from the association. Mutualism is usually obligatory, since in most cases physiological dependence has evolved to such a degree that one mutual cannot survive without the other. Blood-sucking leeches cannot digest blood, and overcome that by harbouring certain intestinal bacterial species to do the digestion for their hosts. At least 20% of insect species, as well as many mites, spiders, crustaceans, and nematodes, are mutually infected with bacteria of genus Wolbachia. Also, filarial nematodes such Wuchereria bancrofti and Onchocerca volvulus which cause serious human diseases, are mutually infected with Wolbachia, and they can be cured of their bacterial infections by treating patients with antibiotics, but the worms die too.

- Commensalism: in which one partner benefits from the association, but the host is neither helped nor harmed. Commensalism may be facultative, in the sense that the commensal may not be required to participate in an association to survive. Humans harbor several species of commensal protozoans, that colonize in the intestinal tract such as Entamoeba dispar, Entamoeba hartmanni, Entamoeba moshkovskii, Entamoeba polecki, Endolimax nana, Iodomoeba butschlii.
- Parasitism: in which one of the participants, the parasite, either harms or lives at the expense of the host. Parasites may cause mechanical injury, such as boring a hole into the host or digging into its skin or other tissues, stimulate a damaging inflammatory or immune response. Most parasites inflict a combination of these conditions on their hosts.

Parasites are different from predators and parasitoids (which also derive benefits from certain interspecific interactions while harming the other participant) in that the host of a parasite is not necessarily killed. Instead, parasites derive benefits from their hosts, most often nutritional resources and shelter, over a longer period of time. It is in fact advantageous to parasites if they do not harm their hosts too badly, because that prolongs the period during which parasites can obtain benefits from hosts. However, in some cases, the impact of parasites on a host is great enough to cause disease, and in extreme cases, the death of the host may also occur.

Classification and General Characters of Human Parasites

The classification of parasites is controversial as there is no universally accepted system.

Parasites form part of the animal kingdom which comprises about 800,000 identified species categorized into 33 phyla.

A Trematode worm, Fasciola hepatica.

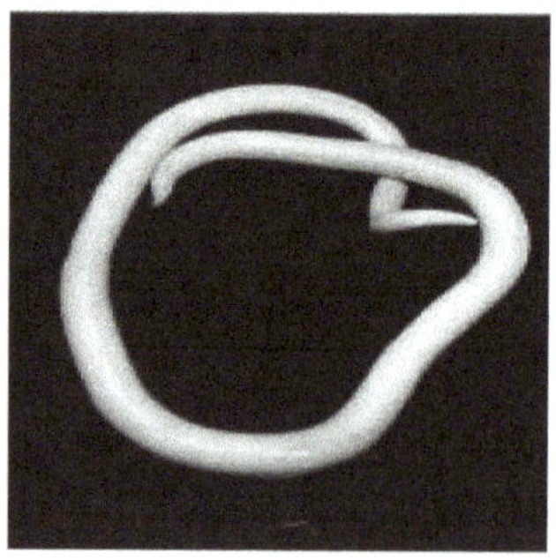

A Nematode worm, Ascaris lumbricoides.

The most acceptable taxonomic classification of human parasites includes Endoparasites and Ectoparasites. Endoparasites are sub-classified into Helminthic parasites (multicellular organisms) and Protozoan parasites (unicellular organisms). Helminthic parasites are either flat worms (Trematodes), segmented ribbon like worms (Cestodes) or cylindrical worms (Nematodes).

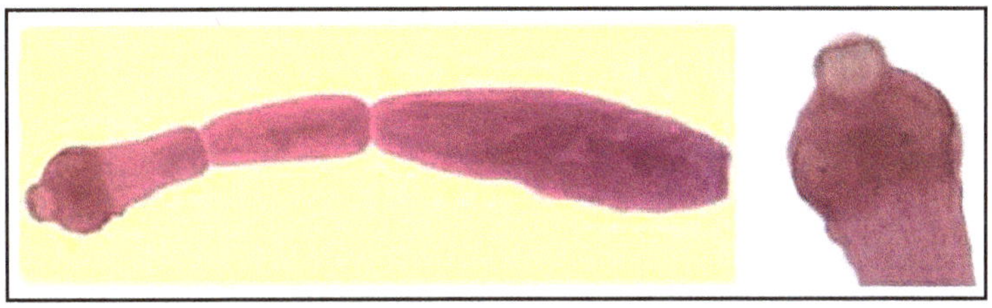

A Cestode worm, Echinococcus granulosus.

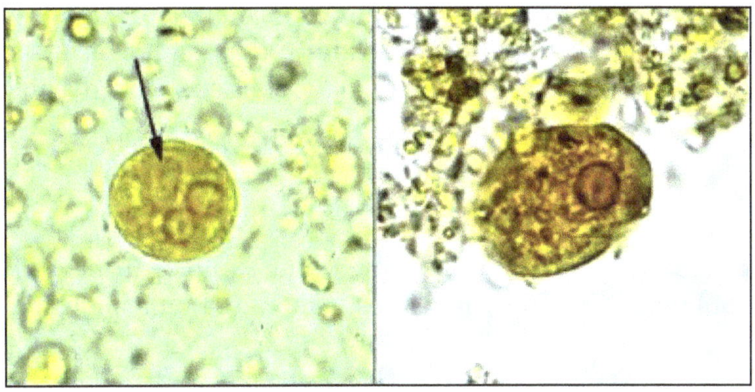

A Protozoan parasite, Entamoeba histolytica (cyst and trophozoite).

Veterinary Parasitology

Veterinary parasitology is the study of animal parasites, especially relationships between parasites and animal hosts. Parasites of domestic animals, (livestock and pet animals), as well as wildlife animals are considered. Veterinary parasitologists study the genesis and development of parasitoses in animal hosts, as well as the taxonomy and systematics of parasites, including the morphology, life cycles, and living needs of parasites in the environment and in animal hosts. Using a variety of research methods, they diagnose, treat, and prevent animal parasitoses. Data obtained from parasitological research in animals helps in veterinary practice and improves animal breeding. The major goal of veterinary parasitology is to protect animals and improve their health, but because a number of animal parasites are transmitted to humans, veterinary parasitology is also important for public health.

Diagnostic Methods

Various methods are used to identify parasites in animals, using feces, blood, and tissue samples from the host animal.

Coprological

Coprological examinations involve examining the feces of animals to identify and count parasite eggs. Some common methods include fecal flotation and sedimentation to separate eggs from fecal matter. Others include the McMaster method, which uses a special two-chamber slide that allows parasite eggs to be more clearly visible and easily counted. It is most commonly used to monitor parasites in horses and other grazing and livestock animals. The Baermann method is similar but requires more specialized equipment and more time and is typically used to diagnose lungworm and threadworm.

Haematological

Haematological examinations involve examining the blood of animals to determine the presence of parasites. Blood parasites tend to inhabit the erythrocytes or white blood cells and are most likely to be detected during the acute phase of infection. Veterinary parasitologists use blood smears, which involve placing a drop of blood onto a slide and spreading it over the surface in a thin film in order to examine it under a microscope. The blood is stained with a dye in order for the cells to be easily distinguished.

Histopathological

Histopathological examinations involve examining tissue samples from animals. A small slice of the organ suspected of being infected by parasites is mounted on a slide, stained, and examined under a microscope.

Though not technically considered a histopathological technique, skin scraping – which involves taking a small sample of the epidermal cells of a dog, cat, or other household pet – is commonly used to detect the presence of mites.

Immunological

Immunological examinations, such as indirect immunofluorescence, ELISA, Immunoblotting (Western blot), and Complement fixation test are methods of identifying different kinds of parasites by detecting the presence of their antigens on or within the parasite itself. These diagnostic methods are used in conjunction with coprological examinations for more specific identification of different parasite species in fecal samples.

Molecular Biological

Molecular biological methods involve studying the DNA of the parasite in order to identify it. PCR and RFLP are used to detect and amplify parasite DNA found in the feces, blood, or tissue of the host. These techniques are very sensitive, which is useful for diagnosing parasites even when they are present in very low numbers; they are also useful for identifying parasites not only in large animal hosts but smaller insect vectors.

Paleoparasitology

Paleoparasitology is the study of parasites from the past, and their interactions with hosts and vectors; it is a subfield of Paleontology, the study of living organisms from the past. Some authors define this term more narrowly, as "Paleoparasitology is the study of parasites in archaeological material." K.J. Reinhard suggests that the term "archaeoparasitology" be applied to "all parasitological remains excavated from archaeological contexts derived from human activity" and that "the term 'paleoparasitology' be applied to studies of nonhuman, paleontological material."

Sources of Material

The primary sources of paleoparasitological material include mummified tissues, coprolites (fossilised dung) from mammals or dinosaurs, fossils, and amber inclusions. Hair, skins, and feathers also yield ectoparasite remains. Some archaeological artifacts document the presence of animal parasites. One example is the depiction of what appear to be mites in the ear of a "hyaena-like" animal in a tomb painting from ancient Thebes.

Some parasites leave marks or traces (ichnofossils) on host remains, which persist in the fossil record in the absence of structural remains of the parasite. Parasitic ichnofossils include plant remains which exhibit characteristic signs of parasitic insect infestation, such as galls or leaf mines and certain anomalies seen in invertebrate endoskeletal remains.

Plant and animal parasites have been found in samples from a broad spectrum of geological periods, including the Holocene (samples over 10,000 years old), Pleistocene (over 550,000 years old), Eocene (over 44 million years old), Cretaceous (over 100 million years) and even Lower Cambrian (over 500 million years).

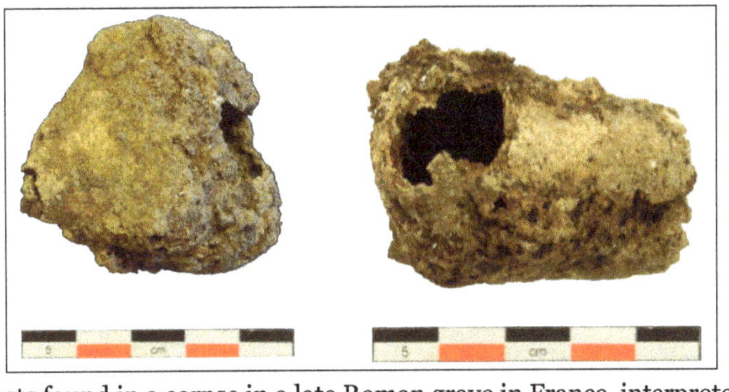

Cysts found in a corpse in a late Roman grave in France, interpreted as signs of probable hydatidosis and capillariasis.

Evidence of Parasitism

One of the most daunting tasks involved in studying parasitic relationships from the past is supporting the assertion that the relationship between two organisms is indeed parasitic. Organisms living in "close association" with each other may exhibit one of several different types of trophic relationships, such as parasitism, mutualism, and commensalism. Demonstration of true parasitism between existing species typically involves observing the harmful effects of parasites on a presumed host. Experimental infection of the presumed host, followed by recovery of viable parasites from that host also supports any claim of true parasitism. Obviously such experiments are not possible with specimens of extinct organisms found in paleontological contexts.

Assumptions of true parasitism in paleontological settings which are based on analogy to known present-day parasitic relationships may not be valid, due to host-specificity. For example, Trypanosoma brucei gambiense and Trypanosoma brucei rhodesiense are both devastating human parasites, but the related subspecies Trypanosoma brucei brucei will infect a number of animal hosts, but cannot even survive in the human blood stream, much less reproduce and infect a human host. So a related (or unidentifiable) species of Trypanosoma found in a paleontological or archaeological context may not be a true human parasite, even though it appears identical (or very similar) to the modern parasitic forms.

The most convincing evidence of paleoparasitism is obtained when a presumed parasite is found in direct association with its presumed host, in a context that is consistent with known host-parasite associations. Some examples include helminths caught in amber in the process of escaping from the body of an insect, lice found in the fur of guinea pig mummies, protozoans in the alimentary canal of flies in amber, nematode larvae found embedded in animal coprolites, and a mite caught in amber in the process of apparently feeding on a spider.

Fossil organisms which are related to present-day parasites often possess the morphological features associated with a parasitic lifestyle, such as blood-feeding mouthparts. So fossil ticks and hematophagous insects are generally assumed to be ectoparasites, even when their remains are found in the absence of a host.

The presence of structures resembling leaf miner trails in leaf fossils provide indirect evidence of parasitism, even if remains of the parasite are not recovered. The dramatic tissue aberrations seen in present-day plant galls and gall-like structures in some invertebrates are direct physiological reactions to the presence of either metazoan parasites or microbial pathogens. Similar structures seen in fossil plant and invertebrate remains are often interpreted as evidence of paleoparasitism.

Host-parasite interactions today are often exploited by other species, and similar examples have been found in the fossil record of plant galls and leaf mines. For example, there are species of wasps, called inquilines, which are unable to induce their own plant galls, so they simply take up residence in the galls that are made by other wasps. Another example is the predation of plant galls or leaf mines, to eat the trapped insect larva inside the gall or mine.

Animal and Plant Parasites

Studies of parasite remains and traces from the past have yielded a vast catalog of ancient host-parasite associations. Genetic sequence data obtained directly from ancient animal parasites, and inferences of past relationships based on genetic sequences of existing parasite groups are also being applied to paleoparasitological questions. Data obtained by all of these methods are constantly improving our understanding of the origin and evolution of the parasites themselves and their vectors, and of the host-parasite and vector-parasite associations.

In some cases, presumed host-parasite relationships of the past seem quite different from those known in the present, such as a fly which appears to be a parasite of a mite.

Paleoparasitological studies have also provided insight into questions outside the realm of parasitology. Examples include the migration and phylogeography of marine mammal hosts, the identity of domestic animal bones based on the known hosts of parasite remains found at the site, and the possible role of climatic changes on animal host genetic diversity.

3
Protozoa

Protozoa are the eukaryotes that feed on organic matter such as other microorganisms, organic tissues and debris. The topics elaborated in this chapter will help in gaining a better perspective about the different types of protozoa such as leishmania, balantidum coli and entamoeba.

Protozoan or Protozoa is an organism, usually single-celled and heterotrophic (using organic carbon as a source of energy), belonging to any of the major lineages of protists and, like most protists, typically microscopic. All protozoans are eukaryotes and therefore possess a "true," or membrane-bound, nucleus. They also are nonfilamentous (in contrast to organisms such as molds, a group of fungi, which have filaments called hyphae) and are confined to moist or aquatic habitats, being ubiquitous in such environments worldwide, from the South Pole to the North Pole. Many are symbionts of other organisms, and some species are parasites.

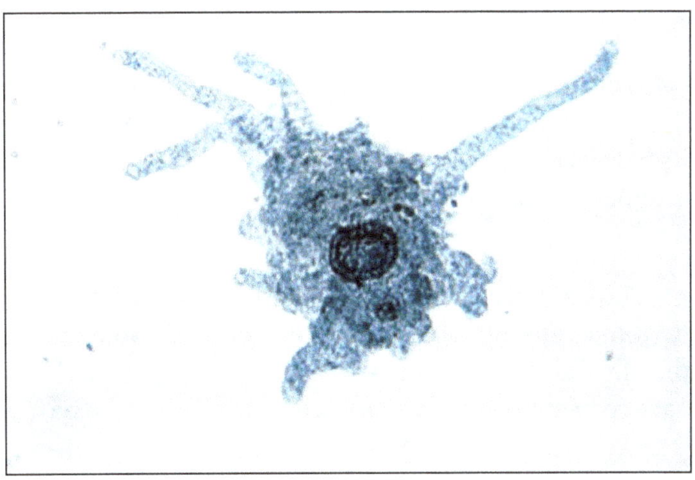

Amoeba (magnified).

Modern ultrastructural, biochemical, and genetic evidence has rendered the term protozoan highly problematic. For example, protozoan historically referred to a protist that has animal-like traits, such as the ability to move through water as though "swimming" like an animal. Protozoans traditionally were thought to be the progenitors of modern animals, but contemporary evidence has revealed that this is not the case for most protozoans. In fact, modern science has shown that the protozoans represent a very complicated grouping of organisms that do not necessarily share a common evolutionary history. This unrelated, or paraphyletic, nature of the protozoans has caused scientists to abandon the term protozoan in formal classification schemes. Hence, the subkingdom Protozoa is now considered obsolete. Today the term protozoan is used informally in reference to nonfilamentous heterotrophic protists.

Commonly known protozoans include representative dinoflagellates, amoebas, paramecia, and the malaria-causing Plasmodium.

Features of Protozoans

Although protozoans are no longer recognized as a formal group in current biological classification systems, protozoan can still be useful as a strictly descriptive term. The protozoans are unified by their heterotrophic mode of nutrition, meaning that these organisms acquire carbon in reduced form from their surrounding environment. However, this is not a unique feature of protozoans. Furthermore, this description is not as straightforward as it seems. For instance, many protists are mixotrophs, capable of both heterotrophy (secondary energy derivation through the consumption of other organisms) and autotrophy (primary energy derivation, such as through the capture of sunlight or metabolism of chemicals in the environment). Examples of protozoan mixotrophs include many chrysophytes. Some protozoans, such as Paramecium bursaria, have developed symbiotic relationships with eukaryotic algae, while the amoeba Paulinella chromatophora remarkably appears to have acquired autotrophy via relatively recent endosymbiosis of a cyanobacterium (a blue-green alga). Hence, many protozoans either perform photosynthesis themselves or benefit from the photosynthetic capabilities of other organisms. Some algal species of protozoans, however, have lost the ability to photosynthesize (e.g., Polytomella species and many dinoflagellates), further complicating the concept of "protozoan."

Protozoans are motile; nearly all possess flagella, cilia, or pseudopodia that allow them to navigate their aqueous habitats. However, this commonality does not represent a unique trait among protozoans; for example, organisms that are clearly not protozoans also produce flagella at various stages in their life cycles (e.g., most brown algae). Protozoans are also strictly non-multicellular and exist as either solitary cells or cell colonies. Nevertheless, some colonial organisms (e.g., Dictyostelium discoideum, supergroup Amoebozoa) exhibit high levels of cell specialization that border on multicellularity.

The descriptive guidelines presented above exclude many organisms, such as flagellated photosynthetic taxa (formerly Phytomastigophora), that were considered protozoans by older classification schemes. Organisms that fit the contemporary definition of a protozoan are found in all major groups of protists that are recognized by protistologists, reflecting the paraphyletic nature of protozoans.

The most important groups of free-living protozoans are found within several major evolutionary clusters of protists, including the ciliates (supergroup Chromalveolata), the lobose amoebae (supergroup Amoebozoa), the filose amoebae (supergroup Rhizaria), the cryptomonads (supergroup Chromalveolata), the excavates (supergroup Excavata), the opisthokonts (supergroup Opisthokonta), and the euglenids (Euglenozoa). These groups of organisms are important ecologically for their role in microbial nutrient cycles and are found in a wide variety of environments, from terrestrial soils to freshwater and marine habitats to aquatic sediments and sea ice. Significant protozoan parasites include representatives from Apicomplexa (supergroup Chromalveolata) and the trypanosomes (Euglenozoa). Organisms from these groups are the causative agents of human diseases such as malaria and African sleeping sickness. Owing to the prevalence of these human pathogens, and to the ecological importance of the free-living protozoan groups mentioned above, much is known about these groups.

Size Range and Diversity of Structure

Protozoans range in diameter from a few thousandths of a millimetre to several millimetres. Because the group contains many unrelated or loosely related organisms, enormous diversity in structure and form exists.

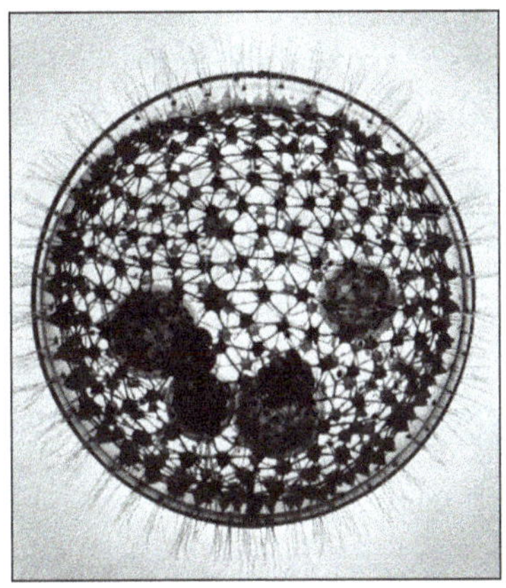

Glass model of protozoan colony Volvox (magnified about 40×).

Flagellated Protozoans

The flagellated protozoans range from a simple oval cell with one or more flagella to the structural sophistication of the collared flagellates (choanoflagellates, supergroup Opisthokonta). The collared flagellates lack photosynthetic pigments and are therefore colourless. They have a single flagellum surrounded by a delicate circular collar of fine pseudopodia (microvilli) on which they trap food particles. In some marine species the whole cell is enclosed in an elaborate, open lattice-like basket formed from strands of silica. Although some dinoflagellates (supergroup Chromalveolata) still contain plant pigments and rely to a greater or lesser degree on photosynthesis, many members have lost the ability to photosynthesize. All dinoflagellates are surrounded by a cell wall armour with a complicated pattern and possess two flagella, one of which beats in a transverse plane around the equator of the cell while the other beats in a longitudinal plane.

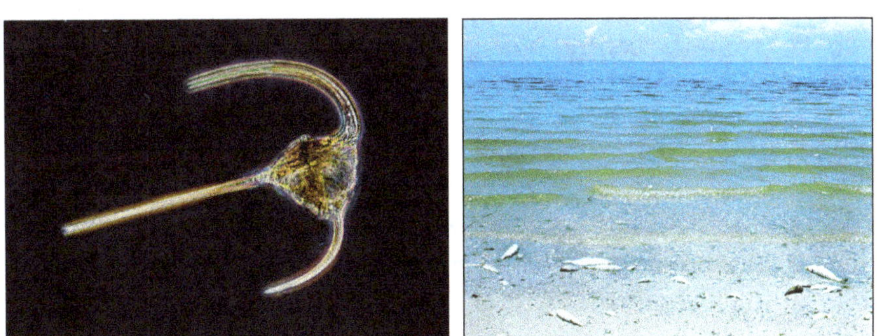

Red tide, Tampa Bay, Florida, showing fish kill and red coloration caused by dinoflagellates.

Many other flagellated protozoans can develop stalks that connect them to a substrate, either as single cells (e.g., the genus Paraphysomonas) or as colonies (e.g., the genus Codosiga). Other flagellated taxa may exist as swimming colonies; for example, in Sphaeroeca and Spongomonas many individual flagellated organisms are embedded in an unstalked gelatinous sphere.

Amoebae and Pseudopodia

The amoebae also are extremely diverse. Amoebae are defined based on pseudopodia type: those with thin, or filose, pseudopods, which may be reinforced by stiff microtubule proteins, are classified in the supergroup Rhizaria (e.g., foraminiferans and radiolarians), whereas those with lobose pseudopods, which are blunt and are not reinforced, are classified in the supergroup Amoebozoa. Both groups of amoebae can be "naked" or housed inside a shell, or test, composed of organic or inorganic materials.

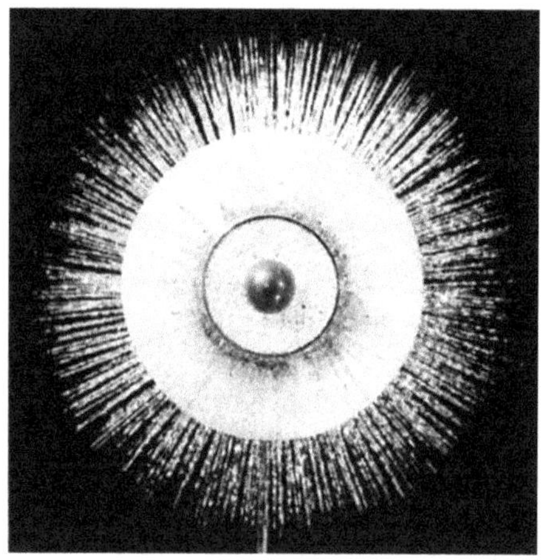

Radiolarian (glass model).

The naked amoebae are the simplest of the amoebae. They have no defined shape and extend one or many lobose pseudopodia. Many of these lobose amoebae, including those in the genera Mastigamoeba and Mastigella, also possess flagella in the vegetative (resting) phase. At the opposite extreme are the complex foraminiferans, which live inside multichambered calcareous shells up to several millimetres in diameter. The filose pseudopodia of foraminiferans are known as reticulopodia and extend from the aperture of the largest chamber of the shell, forming a complicated, sticky branching network. Rhizarian amoebae that are known commonly as radiolarians form shells from silica or strontium sulfate; in some the shell has so many holes that the structure resembles a sponge. The polyphyletic heliozoans, or sun protozoans, have radiating pseudopodia (axopodia) that extend like spokes from the central body; microtubules support an outer layer of cytoplasm. Many heliozoans are members of Rhizaria; however, some are placed in Chromalveolata.

Ciliated Protozoans

The ciliates are the most structurally homogeneous group, although even they have evolved considerable variation on the cilia-covered cell. In some species (e.g., the hypotrich Euplotes) the cilia are combined to form thick conical structures, called cirri, which the ciliate uses to crawl along surfaces, rather like small limbs. In other species the cilia virtually disappear from the main body of the cell, but the circle of cilia around the mouth becomes well developed (as in the oligotrich

Strombidium and the tintinnid ciliates). The peritrich ciliates have developed stalks and attach to plants and animals as a means of dispersal. Many peritrichs (e.g., Epistylis) form branching colonies.

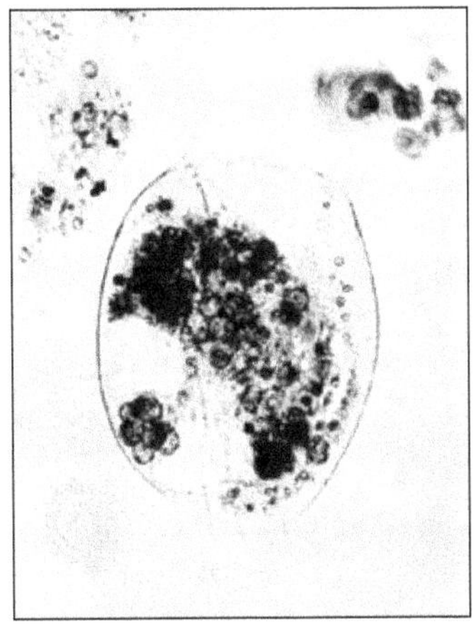

Euplotes patella digesting algae. Movement of cilia has started a whirlpool of algae at right.

The suctorian ciliates have completely lost their cilia in the adult phase. They have instead developed a stalk and many tentacles, which they use to capture passing prey, usually other ciliates. Because they cannot swim, they produce motile ciliated offspring, which settle elsewhere and then transform into the feeding stage, thus avoiding overcrowding.

Parasitic Protozoans

Although the parasitic protozoans tend to be less structurally complex than free-living forms, considerable variation may occur during the course of their life cycles. Plasmodium, the malarial parasite that lives inside the liver and red blood cells of humans and the gut of its insect vector (the Anopheles mosquito), undergoes various changes in form through its asexual and sexual phases of development. Among the parasitic flagellates, the trypanosomes and their relatives (kinetoplastids), morphological variation occurs during the various stages of the life cycle in both the mammalian and insect hosts. Among species of Leishmania, which cause visceral leishmaniasis (kala-azar), cutaneous leishmaniasis (Oriental sore), and mucocutaneous leishmaniasis (espundia), two distinctly different forms occur. Rounded, nonflagellated forms called amastigotes feed and divide inside macrophage cells in different regions of the human body, while in the gut of the insect vector there occurs a flagellated form called a promastigote. Members of the genus Trypanosoma, which cause sleeping sickness and other diseases, have flagellated forms with different morphologies. At some stage in the life cycle, all assume the trypomastigote form—i.e., slender with part of the flagellum running over the body and attached to it by a finlike extension to form an undulating membrane. They may also occur as amastigote (stumpy flagella) or promastigote forms.

Distribution and Abundance

Protozoans have colonized a wide array of aquatic and terrestrial habitats from the Arctic and Antarctic to equatorial zones. In soils and bogs, protozoans form part of a complex microbial community. They live in the moisture films surrounding soil particles, so that they are actually aquatic organisms, even though living in a terrestrial environment. Between 10,000 and 100,000 organisms per gram of soil may inhabit fertile land; the relative proportions of each group vary depending on soil type and latitude. In Antarctic soils flagellates and testate (shell-dwelling) amoebae predominate, while in temperate woodland soils ciliates are more numerous.

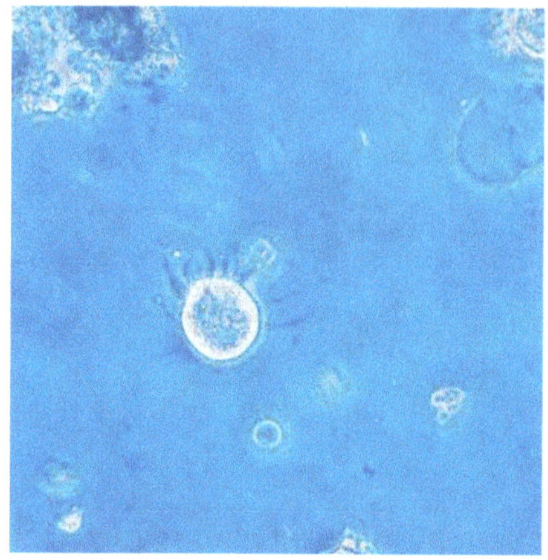

Oligotrich: Halteria grandinella.

In the open waters of lakes, estuaries, and the ocean, protozoans form an important component of the floating (planktonic) community. They are often present in densities of tens of thousands per litre of water. Most planktonic protozoa feed on bacteria, algae, other protozoans, and small animals. The most common planktonic protozoans include a variety of flagellated taxa, ciliates—especially oligotrichs and tintinnids (which live inside small tubes, or loricae)—and the exclusively marine foraminiferans and radiolarians. Foraminiferans have been found at depths of 4,000 metres (about 13,120 feet), and some protozoans have been observed around hydrothermal vents on the ocean floor.

Ecological and Industrial Importance of Protozoans

Protozoans play important roles in the fertility of soils. By grazing on soil bacteria, they regulate bacterial populations and maintain them in a state of physiological youth—i.e., in the active growing phase. This enhances the rates at which bacteria decompose dead organic matter. Protozoans also excrete nitrogen and phosphorus, in the form of ammonium and orthophosphate, as products of their metabolism, and studies have shown that the presence of protozoans in soils enhances plant growth.

Protozoans play important roles in wastewater treatment processes, in both activated sludge and slow percolating filter plants. In both processes, after solid wastes are removed from the sewage, the remaining liquid is mixed with the final sludge product, aerated, and oxidized by aerobic microorganisms to consume the organic wastes suspended in the fluid. In the activated sludge process, aerobic ciliates consume aerobic bacteria, which have flocculated (formed loose aggregates, making them easily separated from liquid). In the percolating filter process, substrates are steeped in microorganisms, such as fungi, algae, and bacteria, which provide food for oxidizing protozoans. In the final stages of both processes, solids settle out of the cleaned effluent in the settlement tank. Treatment plants with no ciliates and only small numbers of amoebae and flagellates produce turbid effluents containing high levels of bacteria and suspended solids. Good-quality, clean effluents are produced in the presence of large ciliated protozoan communities because they graze voraciously on dispersed bacteria and because they have the ability to flocculate suspended particulate matter and bacteria.

Protozoans probably play a similar role in polluted natural ecosystems. Indeed, there is evidence that they, by feeding on oil-degrading bacteria, increase bacterial growth in much the same way that they enhance rates of decomposition in soils, thereby speeding up the breakdown of oil spillages.

Some radiolarians and foraminiferans harbour symbiotic algae that provide their protozoan hosts with a portion of the products of photosynthesis. The protozoans reciprocate by providing shelter and carbon and essential phytonutrients. Many ciliates contain endosymbiotic algae, and one species, Mesodinium rubrum, has formed such a successful relationship with its red-pigmented algal symbiont that it has lost the ability to feed and relies entirely on symbiosis for its livelihood. Mesodinium often forms dense red blooms, or red tides, when it reaches high densities in water. Among the ciliates with endosymbionts, Mesodinium is the only completely photosynthetic species. Other ciliates achieve photosynthesis in another way. Although they do not have symbiotic algae, they consume plantlike flagellates, sequester the organelles that contain the plant pigments, and use them for photosynthesis. These organelles are known as plastids. Because the isolated plastids eventually age and die, they must be replaced continuously.

The impact of protozoan grazing on phytoplankton can be considerable. It has been estimated that at least half of the phytoplankton production in marine waters is consumed by protozoans. Like the soil protozoans, these planktonic protozoans excrete nitrogen and phosphorus at high rates. The protozoans are a fundamental component in recycling essential nutrients (nitrogen and phosphorus) to the phytoplankton.

Protozoans and Disease

Parasitic protozoans have invaded and successfully established themselves in hosts from practically every animal phylum. The best-studied parasitic species are those of medical and agricultural relevance. The trypanosomes, for example, cause a number of important diseases in humans. African sleeping sickness is produced by two subspecies of Trypanosoma brucei—namely, T. brucei gambiense and T. brucei rhodesiense. The life cycle of T. brucei has two hosts: a human (or other mammal) and the bloodsucking tsetse fly, which transmits the parasite between humans.

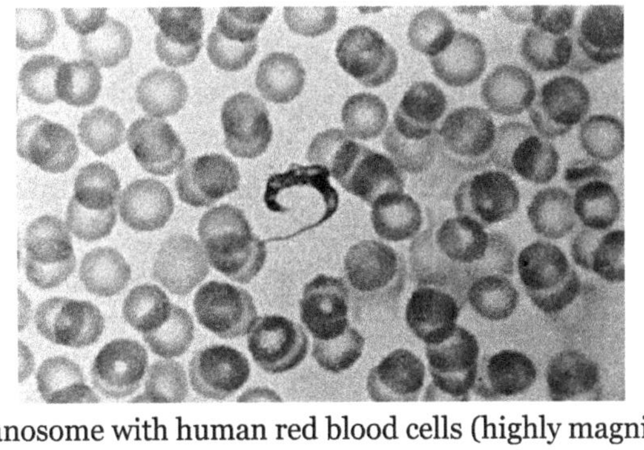

Trypanosome with human red blood cells (highly magnified).

Trypanosomes live in the blood plasma and the central nervous system of humans and have evolved an ingenious way of fooling the immune system of the host. Upon contact with a parasite, the immune system generates antibodies that recognize the specific chemical and physical nature of the parasite and actively neutralize it. Just as the host's immune system is beginning to win the battle against the parasite and the bulk of the population is being recognized and destroyed by host antibodies, the parasite is able to shed its glycoprotein coat, which is attached to the cell surface, and replace it with a coat containing different amino acid sequences. Thus, the parasite essentially changes its makeup. These alternate forms are known as antigenic variants, and it has been estimated that each species may have as many as 100 to 1,000 such variants. The host must produce a new set of antibodies against each new variant, and in the meantime the parasite has time to replenish its numbers. Ultimately, unless the disease is treated, the parasite wins the battle and the host dies. Such antigenic variation makes the development of an effective vaccine against certain parasitic protozoan diseases virtually impossible.

A close relative of T. brucei, Trypanosoma cruzi, causes Chagas disease, or American trypanosomiasis. Vector hosts include bugs of the genus Rhodnius and other arthropods, such as lice and bedbugs. In humans the nonflagellated (amastigote) form of the parasite lives inside macrophage cells, the cells of the central nervous system, and muscle tissue, including the heart, where it grows and divides. Short trypomastigote flagellated forms periodically appear in the blood, where they are readily taken up by the bloodsucking vector hosts. These flagellated forms do not divide in the blood; reproduction occurs only in the amastigote intracellular forms.

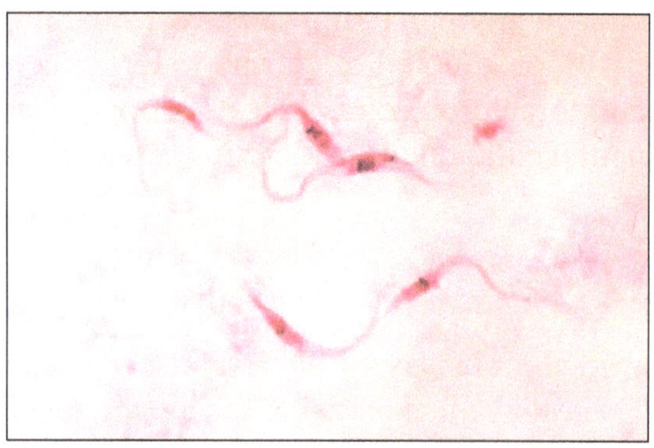

Photomicrograph of Trypanosoma cruzi, the causative agent of Chagas disease.

Relatives of the trypanosomes, species of the genus Leishmania, cause a variety of diseases worldwide, known as leishmaniasis. Like T. cruzi, these are intracellular parasites of the macrophage cells. The intermediate, or vector, hosts are a variety of sand fly species (subfamily Phlebotominae). In cutaneous leishmaniasis the infected macrophages remain localized at the site of the infection, causing an unsightly lesion, but in visceral leishmaniasis the infected macrophages are carried by the blood to the visceral organs. This latter disease is characterized by enlargement of the spleen and liver, leading to the distended abdomen that is typical of kala-azar. In mucocutaneous leishmaniasis the initial skin infection spreads to the mucous membranes of the face (the nose, mouth, and throat), producing a lesion that can cause destruction of part of the face.

Malaria, which is caused by the apicomplexan protozoan Plasmodium, remains a serious disease despite measures that can be taken to control and eradicate the mosquito vector host and despite the availability of an array of antimalarial drugs. The life cycle is fundamentally identical among the five species of Plasmodium, but the pathology of the disease varies in the frequency and severity of attacks and in the occurrence of relapses. Problems in controlling the disease include the development of resistance to insecticides by the mosquito and the evolution of drug resistance by the parasite. Prophylactic drugs taken before and during a visit to areas where malaria is endemic may prevent the disease from forming in persons who have no natural resistance. Antigenic variation does not appear to occur in Plasmodium, which is promising for vaccine development.

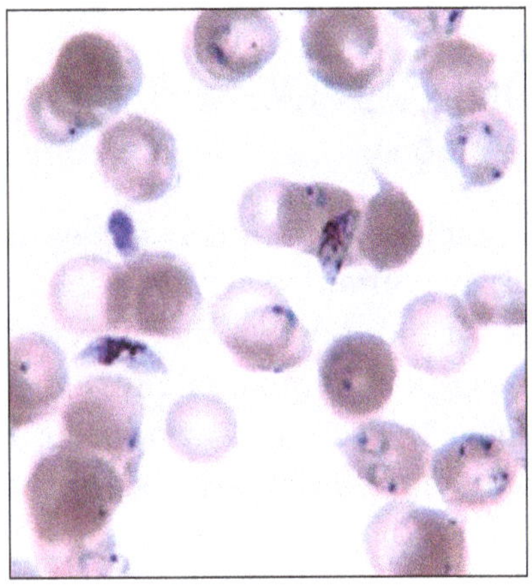

Blackwater Fever.

Micrograph of blood cells showing ring-forms (circular organisms within the cells) and gametocytes (purplish oblong shapes) of Plasmodium falciparum.

The apicomplexan Cryptosporidium is a protozoan parasite of humans and other mammals that was discovered in the 1970s. It has a one-host life cycle and lives inside the cells lining the intestines and sometimes the lungs. Cryptosporidium carries out all the asexual reproductive stages typical of an apicomplexan inside a single host and is passed from host to host in a resistant cyst stage called an oocyst. The disease caused by the parasite is typified by severe diarrhea and vomiting. Although there is no drug treatment, most healthy people recover quickly. In persons who have impaired immune systems, such as AIDS patients, however, Cryptosporidium can cause serious infections.

Form and Function

Protozoan Cell

The protozoan cell carries out all of the processes—including feeding, growth, reproduction, excretion, and movement—necessary to sustain and propagate life. The cell is enclosed in a membrane called the plasma membrane. Like all membranous structures in the eukaryotic cell, the plasma membrane is composed of mostly lipid and some protein molecules. The plasma membrane is a barrier between the cell cytoplasm and the outside liquid environment. Some substances, such as oxygen, readily pass through the membrane by diffusion (passive transport), while others must be transported across at the expense of energy (active transport). Cilia and flagella arising from the cell are also sheathed in the cell membrane; this is in contrast to bacterial flagella, which are not surrounded by a membrane.

The cell also has internal membranes, which are not as thick as the plasma membrane. Among these are the endoplasmic reticulum, whose membranes separate compartments of the cell, thereby allowing different conditions to be maintained in various parts—e.g., separation of deleteriously reactive substances. Enzymes are arranged on the surface of the endoplasmic reticulum; one such enzyme system catalyzes the activity of the ribosomes during protein synthesis. The Golgi apparatus is a cluster of flattened vesicles, or cisternae, associated with the endoplasmic reticulum. The vesicles are involved in membrane maturation and the formation and storage of the products of cell synthesis, as in the formation of scales on the surface coat of some flagellates, for example. The scales are formed within the Golgi and are transported by the vesicles to the plasma membrane, where they are incorporated onto the surface of the cell. The Golgi apparatus is poorly evident in most ciliates and absent from some amoebae.

All protozoans possess at least one nucleus, and many species are multinucleate. The genetic material DNA (deoxyribonucleic acid) is contained within the chromosomes of the nucleus. Each nucleus is bounded by two unit membranes possessing pores that permit the passage of molecules between the cytoplasm and the nucleoplasm. Most ciliates have two types of nuclei: micronuclei and macronuclei. The macronucleus is the somatic, or nonreproductive, nucleus. It is large and it is polyploid, meaning that it contains more than two sets of chromosomes (the condition of two sets of chromosomes is described as diploid). In contrast, the micronucleus is germinal (responsible for transfer of genetic information during sexual reproduction) and diploid. The macronucleus can be quite variable in shape, resembling in some species a string of beads or a horseshoe. It directs the normal functioning of the cell and usually disintegrates during sexual reproduction, to be re-formed from the products of micronuclear division after the sexual phase is completed.

Almost all protozoans contain double-membrane mitochondria; the inner membrane forms flattened, tubular, or discoidal extensions (cristae) into the mitochondrial interior in order to increase the surface area of the respiratory machinery, and the outer membrane forms the boundary of the organelle. Mitochondria are the sites of cellular respiration in most eukaryotes. Species that do not require oxygen (anaerobes), such as those that live in the intestinal tract of their hosts or those that occupy special anaerobic ecological niches, lack mitochondria. Instead, they have energy-generating organelles, such as hydrogenosomes and mitosomes, that belong to the family of organelles called microbodies. These oblong or spherical membrane-bound organelles, about 1–2 micrometres (μm; 1 micrometre = 3.9 × 10–5 inch) in length, are believed to be the site of fermentative processes. They contain enzymes that oxidize pyruvate to acetate and carbon dioxide, resulting in the release of hydrogen sulfide under anaerobic conditions.

Organisms that live in a liquid environment with a lower concentration of ions than is found in the interior of their cells—an osmotically hypotonic environment—gradually gain water if they equilibrate with their habitat. If this process remains unchecked, the cell swells and bursts. In protozoans the maintenance of the osmotic gradient between the cell cytoplasm and the environment is achieved by the contractile vacuole. These membrane-bound organelles are situated close to the plasma membrane. They swell with water periodically and then suddenly contract and disappear, forcing their contents from the cell in repeated cycles. In some amoebae and some flagellated taxa the contractile vacuole is formed when smaller vesicles combine with the main vacuole. In the ciliates the contractile vacuole is fed by a complex system of feeder canals, which are in turn fed by a complex network of vesicles and fine tubules within the cytoplasm.

Protozoans have transitory food or digestive vacuoles. The number of these membrane-bound cell organelles depends on the feeding habits of the organism. Some species may have many, whereas others may contain only one or two at any one time. In ciliates the food vacuoles form at the base of the cytopharynx, whereas in species without a cell "mouth," or cytostome, the vacuoles form near the cell membrane at the site where food is ingested.

Within the cell, structural proteins of various types form the cytoskeleton (cell skeleton) and the locomotory appendages. They include microfilaments formed of a contractile protein also found in the muscles of animals (actin) and cylindrical microtubules formed from filaments of the protein tubulin. Microtubules are particularly important in the structural formation and functioning of cilia and flagella. Filopodia of certain rhizarian species are supported by microtubules.

Characteristics of Locomotion

Protozoans exhibit diverse modes of locomotion across the various groups, but the modes of locomotion can be broadly divided into flagellar, ciliary, and amoeboid movement. Only the ciliates among the three major motility groups of protozoans, however, represent a truly monophyletic group (or single evolutionary line). (Some non-ciliates, such as those of group Opalinata, possess cilia-like organelles that are fundamentally different from true cilia.) In contrast, flagella and pseudopodia are present in a wide variety of distantly related taxa.

Flagellar Propulsion

Flagellar propulsion is employed during some stages in the life cycles of certain amoebae, including the vegetative phase of some genera, such as Mastigamoeba and Mastigella. The eukaryotic flagellum is a membrane-bound, whiplike structure found not only in protozoans but in animals as well (such as in sperm, the male reproductive cells of animals). The structure of the eukaryotic flagellum consists of a cylinder (axoneme) made up of a pair of central microtubules surrounded and joined by cross-bridges to a circle of nine pairs of microtubules. This "nine-plus-two" arrangement of the microtubules in the axoneme is surrounded by cytoplasm and ensheathed in cell membrane. The flagellum arises from the basal body, or kinetosome, within the cell.

The undulating motion of the flagellum is normally generated at its base. The waves move along the flagellum to produce a force on the water acting along the long axis of the organelle in the direction of the wave. The speed of movement is determined by the length of the flagellum and by the size of, and distance between, the waves it generates. Species of monophyletic stramenopiles (heterokonts) have tripartate tubular hairs (mastigonemes) arising at right angles to the flagellum along its length, whereas other groups, such as the dinoflagellates and euglenids, have slender, simpler hairs called flimmer filaments. Either structure improves the effectiveness of the flagellar stroke, altering the movement of water produced by undulations of the flagellum by reversing its flow toward the flagellar base. Swimming speeds achieved by flagella are relatively low.

Cilium Structure and Beat

Ciliates have an increased number of beating flagella on the cell surface, thereby enabling greater power and speeds to be developed against viscous forces. The structure of a cilium is identical to that of a flagellum, but the cilium is considerably shorter. Cilia are a type of flagella arranged in closely aligned longitudinal rows called kineties. A complex system of fibres and microtubules arising from the basal bodies, or kinetosomes, of each cilium connects it to its neighbouring cilia in the kinety and to adjacent ciliary rows. In some species the body cilia may be reduced to specialized cirri, where the kinetosomes are not arranged in the usual rows but instead have a hexagonal pattern interlinked at several levels by fibres and microtubules.

The effective stroke of the cilium is usually planar, but in the recovery stroke the cilium sweeps out to the side, creating an overall beat with a three-dimensional pattern. The cilium performs work against the viscous force of the water during both the effective and the recovery strokes. To be effective, each cilium must beat in a coordinated manner with its neighbouring cilia. A synchronized beat is passed along a ciliary row by means of a hydrodynamic linkage between the cilia. During a beat, each cilium displaces a layer of surrounding water. Displaced water layers overlap between cilia and, as a consequence, interference occurs between the movements of adjacent cilia, creating a hydrodynamic linkage.

Amoeboid Movement

Amoeboid movement is achieved by pseudopodia and involves the flow of cytoplasm as extensions of the organism. The process is visible under the light microscope as a movement of granules within the organism. The basic locomotory organelle is the pseudopodium. The way in which movement is effected can vary slightly among groups but generally involves the polymerization of cytoskeletal proteins (actin and myosin) at the leading edge of the pseudopod, followed by the flow of cytoplasmic material into the vacancy produced through the polymerization process. The flow of cytoplasm provides the momentum necessary to propel the organism further in its direction of movement. Additional forces driving the amoeboid movement involve the "eupodium," which extends into a potential substrate for a grab-like traction, similar to a tank tread. Pushing force is also generated in the posterior end of the organism by contractions of the cytoskeletal proteins.

A variety of pseudopodial types are found among the naked and testate amoebae. In some species a single pseudopodium is extended at any one time; in others, numerous tubular pseudopodia are extended simultaneously. Some amoebae appear saclike throughout locomotion, and no pseudopodia are obvious. The numerous long, stiff protoplasmic extensions (axopodia) of the amoeboid (paraphyletic) heliozoans shorten and lengthen—the forward axopodia lengthen and become attached, while the posterior axopodia detach and retract—and the amoeba rolls slowly along. The foraminiferans move by extending slender pseudopodia (filapodia), which may be several millimetres long in some species. The extending filopodia branch and fuse with each other so that there is a complex, continuously changing network of pseudopodia pulling the organism along.

Respiration and other Energy-generating Pathways

Aerobic Protozoans

Most species of free-living protozoans appear to be obligate aerobes (they cannot survive without oxygen). As in the cells of animals, plants, and fungi, their respiration is based on oxidation (with molecular oxygen, O_2) of the six-carbon glucose molecule, resulting in the formation of carbon dioxide molecules and water. In protozoans and eukaryotes in general, metabolism and respiration occur stepwise via three specific pathways: the Embden-Meyerhof-Parnas pathway (glycolysis), the tricarboxylic acid cycle (also known as the Krebs cycle, or citric acid cycle), and the electron transport chain, which uses cytochromes, flavins, and quinones as electron carriers. In some protozoans (and in nearly all other eukaryotes) the last two metabolic processes (the tricarboxylic acid cycle and electron transport) take place in mitochondria.

Aerobic protozoans are so small that they are able to obtain the oxygen they require for metabolism from the surrounding liquid medium by simple diffusion. The special pigments or structures required for the acquisition and transport of oxygen that are found in multicellular organisms are not required in protozoans. The pigment hemoglobin has been found in some ciliates (e.g., Tetrahymena), although it does not appear to function as an oxygen-carrying pigment as it does in humans. Within a single species, the rate of oxygen consumption varies in relation to factors such as temperature, the stage in the life cycle, and the cell's nutritional status (i.e., whether or not it is well fed).

Anaerobic Protozoans

Obligate anaerobes, in which metabolism must take place in the absence of oxygen, are rarely found among eukaryotic organisms. Those eukaryotes that are anaerobic often are either parasites or obligate symbionts of multicellular organisms that have evolved from aerobic ancestors. Excavata includes several anaerobic groups; many of its parasitic and symbiotic taxa live in the gastrointestinal tracts of invertebrates and vertebrates, such as humans. For instance, the diplomonad Giardia is an anaerobic parasite found in contaminated water that causes the gastrointestinal disease giardiasis. Trichomonads are a large group of anaerobic parasites. The organism Trichomonas vaginalis causes the sexually transmitted disease trichomoniasis. Trichomonads are closely related to the hypermastigotes, a group of anaerobes that are obligate symbionts of wood-digesting insects. Another large group of anaerobic symbionts of wood-digesting insects are the oxymonads. Some anaerobic protozoans are free-living.

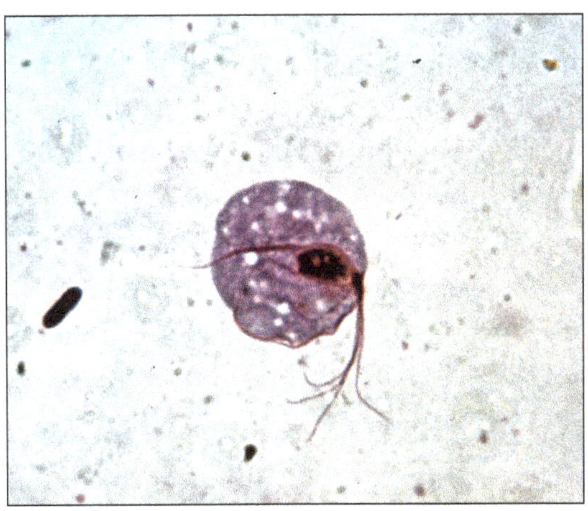

Trichomonas vaginalis.

One ecological group of ciliates (e.g., Metopus, Plagiopyla, and Caenomorpha) is associated with sulfide-containing sediments. The sulfur ciliates harbour endosymbiotic and ectosymbiotic bacteria, which may take the metabolic end products released by the ciliates and reutilize them for growth and energy-yielding processes. Similar to other anaerobic protozoans, these ciliates are believed to have reverted from an aerobic metabolism to an anaerobic lifestyle in order to exploit a specialized ecological niche.

Hydrogenosomes

Unlike typical eukaryotic mitochondria, many anaerobic protozoans possess energy-yielding organelles belonging to a family of cellular structures called microbodies. Thus, these organisms do not perform the tricarboxylic acid cycle, nor do they possess electron transport chains. Instead, they must rely on substrate-level phosphorylation for the generation of the energy molecule adenosine triphosphate (ATP). A microbody commonly found in anaerobic protozoans is the hydrogenosome. Hydrogenosomes are enveloped by a double membrane and generate cellular energy via the partial oxidation of pyruvate to acetate (pyruvate fermentation). This reaction results in the production of carbon dioxide, molecular hydrogen (H_2), and ATP. The principle enzyme, pyruvate ferredoxin oxidoreductase, is present in high concentrations and forms conspicuous crystalline structures inside the hydrogenosome. The hydrogenosome is found in the trichomonads, hypermastigotes, and some euglenids. Hydrogenosomes are thought to have evolved from mitochondria.

Mitosomes and Glycosomes

Another type of anaerobic organelle common among anaerobic protozoans is the mitosome. The mitosome likely evolved from mitochondria, independent of hydrogenosome evolution. Mitosomes are found in diplomonads such as Giardia and were originally described in the intestinal parasite Entamoeba histolytica.

Certain parasitic protozoans that live in the blood, such as Trypanosoma brucei, have evolved a system of energy generation that makes use of yet another type of organelle, the glycosome. The glycosome contains glycolytic enzymes that oxidize glucose to the three-carbon molecule pyruvate. The glycosome is related to the peroxisome, a nearly ubiquitous eukaryotic organelle.

Carbon Acquisition and Nutrition

By definition, protozoans are nonfilamentous heterotrophs, meaning that they acquire carbon in the form of organic carbon from external sources, without the use of absorptive structures that are funguslike (i.e., hyphaelike). Instead, protozoans may ingest organic carbon substrates using phagotrophy. Food sources may include bacteria, algae, other protozoans, and small animals, such as the crustacean copepods.

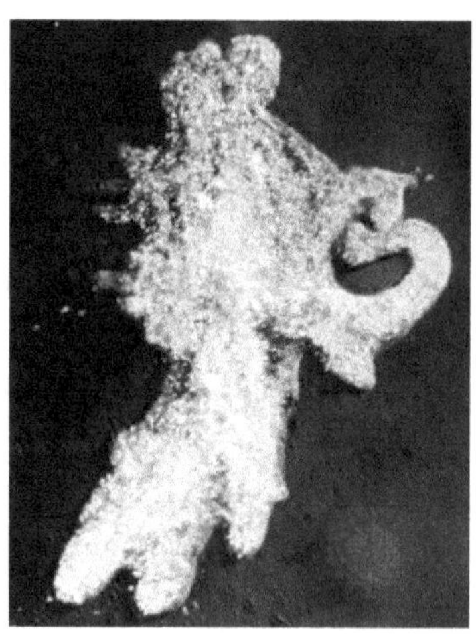

Amoeba engulfing a ciliate.

Mechanisms of Food Ingestion

Protozoans may take food into the cell at a specific point, such as the cytostome (a well-developed feeding groove), at a particular region of the cell surface, or at any random point of entry. In the collared flagellates, or choanoflagellates, for example, the collar and flagellum operate in feeding. The collar, composed of fine pseudopodia, surrounds the flagellum. The beating flagellum creates a water current, causing water to move through the collar. Particles of food in the current are trapped on the collar and are ingested by pseudopodia at its base. The ingested food is then enclosed in a membrane-bound digestive or food vacuole.

Many ciliates are filter feeders, creating water currents with special ciliary structures associated with the cytostome. The synchronized beating of these ciliary structures pushes a stream of water against a membranelle composed of cilia. The membranelle acts as a collecting sieve, where the food particles become trapped in the free spaces between the cilia. Using this mode of feeding, ciliates can shift considerable volumes of water in relation to their size. Tetrahymena, for example, can filter 3,000 to 30,000 times its own volume in one hour.

Other ciliates lack complex oral cilia and gather their food by other means. Nassula has a complex cytostome and cytopharynx supported by a basketlike cytopharyngeal structure composed of microtubules. This species ingests filamentous algae by grasping the filament, bending it like a hairpin, and drawing it into the cytopharynx, where it is broken up into fragments and enclosed in digestive vacuoles. Predatory ciliates such as Didinium nasutum, Lacrymaria olor, and Dileptus anser apprehend their prey with special structures called extrusomes. Among the various types of extrusomes are the toxicysts, which are found in the oral region and release toxins that paralyze the prey.

The suctorians are ciliate predators that usually possess tentacles of two functional types: feeding tentacles and piercing tentacles. The latter trap and immobilize the prey, usually other ciliates that make chance contact with the outstretched tentacles of the suctorian. The cell contents of the prey are transported up through the feeding tentacles into the suctorian, where digestive vacuoles are formed. The transporting mechanism is mediated by a complex array of microtubules within the tentacle. A single suctorian can often feed on several prey at the same time, and frequently the prey are larger than the predator.

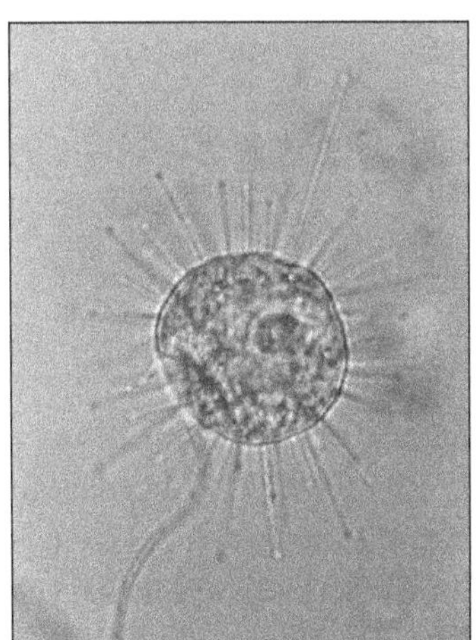

Suctorian: The ciliated suctorian Podophrya fixa.

Amoebae, all of which lack a cell mouth, or cytostome, also exhibit a diverse array of feeding mechanisms and diet. Some feed on filaments of cyanobacteria (blue-green algae)—which are composed of long chains of individual cells—by taking in the entire filament at any point on the cell surface and rolling it up into a coil inside a digestive vacuole. Others, such as the testate amoeba Pontigulasia, pierce single cells in algal filaments and remove the contents. The radiolarians and foraminiferans trap a wide range of prey, including protozoans, algae, and small crustaceans, in their complex pseudopodial networks and then transport the food items along the pseudopodia to the main body of the cell for ingestion. This process gives foraminiferans a distinctive appearance under the light microscope; relatively large grains of food particles are scattered across the reticulate pseudopodia, earning the organisms the name granuloreticulosans.

Parasitic protozoans feed in a variety of ways. Many live in the nutrient-rich medium of the body fluids—e.g., the blood or cells of their host. There they take in energy-rich fluids by pinocytosis, in which small amounts of the medium are pinched off into digestive vacuoles either at a specific site, such as the cytostome in ciliates or the flagellar pocket in trypanosomes, or along the surface of the cell in amoebae. Other parasitic protozoans engulf portions of the host tissue through phagocytosis in much the same way that free-living amoebae feed. Plasmodium, for example, engulfs portions of the red blood cells or liver cells in which they live. The hemoglobin in the cytoplasm of the red blood cell is only partially digested by the parasite. The protein portion of the hemoglobin molecule is degraded to its constituent amino acids, but the iron-containing portion is converted into insoluble iron-containing hemozoin, which remains within the parasite's endosomes until discarded at the next division. This process removes free hematin from the parasite cytoplasm, where it would otherwise prevent further metabolism within the parasite because it inhibits the actions of succinic dehydrogenase, an enzyme in the tricarboxylic acid cycle.

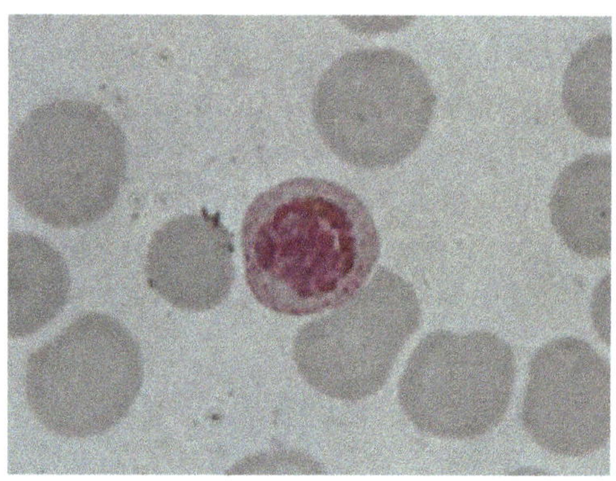

Plasmodium vivax in red blood cell.

Food Vacuoles

Whatever the mode of heterotrophic nutrition or diet, the food material is enclosed in food vacuoles, which are bounded by cell membrane. Digestive enzymes are poured into the newly formed vacuole from the surrounding cytoplasm. In the ciliate Paramecium, where the process has been researched in detail, the digestive vacuoles initially decrease in size and the enclosed particles aggregate. As digestion proceeds, the vacuole increases in size and the contents become progressively acidic, before gradually becoming alkaline near the end of the process. The products of digestion are then absorbed into the surrounding cytoplasm, and the waste material is ejected from the cell anus, or cytoproct. The length of the digestive cycle varies and depends on the species and the diet.

Paramecium contains a reservoir of membrane-forming material in discoid vesicles for the purpose of producing food vacuoles. The food vacuoles form at the cytopharynx when the cytopharyngeal membrane and the discoid vesicles fuse. At the cytoproct, where the vacuoles are broken down and the waste material of digestion is ejected, the membrane material is retrieved and returned to the cytopharynx. Thus, the pool of digestive vacuole membrane is continuously recycled within the cell.

Food Selection

While they seem to lack a sensory system, protozoans are capable of food selection. Many of the filter feeders apparently discriminate solely on the basis of size, dictated by the dimensions of the spaces in the membranelle acting as a sieve. Some filter-feeding ciliates, such as the tintinnids, however, are known to be selective and appear to be able to capture or reject items that arrive at the feeding membranelles in the feeding current. The large ciliate Stentor, for example, takes ciliates in preference to flagellated cells and algae, and discrimination increases as the animal becomes less hungry. Carnivorous species exercise distinct selectivity. Most suctorians feed exclusively on particular ciliate taxa. They are selective feeders and usually do not capture flagellates, amoebae, or their own ciliated swarmers. Evidence suggests that a reaction between chemical compounds on the surface of the prey and the tentacle tip of the suctorian is responsible for feeding selectivity. Amoeboid organisms also display feeding selectivity. Amoeba proteus, for example, selects the flagellate Chilomonas paramecium in preference to Monas punctum, even when the number of Monas in the medium is high. In this case, selection may be based on the digestibility of the prey; the latter is digested in 3 ½ hours, the former in 3 to 18 minutes.

Mixotrophy

All protozoans engage in heterotrophy, but not all protozoans are exclusive heterotrophs. Those that combine autotrophy (self-sustaining food production from a carbon source and inorganic nitrogen) and heterotrophy (ingesting other organisms to acquire carbon) are known as mixotrophs. The degree of mixotrophy in a protozoan varies from complete reliance on the symbiotic alga (or algae) to transitory retention of the plastids of phytoflagellate prey with only a partial dependence on photosynthesis to supplement the cell's energy balance. For example, many protozoans, including the predatory ciliate Stentor and the heliozoan Acanthocystis, are capable of forming ephemeral symbioses with the coccoid green alga Chlorella. The ciliate Paramecium bursaria forms longer-lasting symbiosis with Chlorella but must nevertheless acquire the alga with each new generation.

Photosynthesis and Plastid Acquisition

Many coloured (i.e., photosynthetic) protists combine autotrophy with heterotrophy and therefore are mixotrophs. For example, some members of the euglenid and cryptomonad groups are mixotrophs. The mixotrophic members of these groups are commonly called acetate flagellates because their preferred organic carbon sources are acetates, simple fatty acids, and alcohols. These organisms are able to switch from carbohydrate-producing photosynthesis when light is available to heterotrophy on acetate and other substrates when light is not available.

In another form of mixotrophy, the amoebae and ciliates sequester only the plastids of their algal prey (rather than a complete algal cell) and use the plastids for photosynthesis. The plastids do not replicate inside the protozoan (as they do in the symbiotic algae), and thus they must be replaced continuously. The large marine ciliate Tontonia appendiculariformis may contain thousands of plastids that have been derived from a variety of flagellates. T. appendiculariformis, however, appears to be selective in its choice of prey, deriving plastids only from certain organisms.

Symbiotic Mixotrophy

Mixotrophy is a common phenomenon among free-living protozoans, which usually obtain the capability of photosynthesis from symbionts that are acquired with each new generation (i.e., the photosynthetic machinery is not inherited). Symbiotic mixotrophy, however, is not "true" mixotrophy, since the combined metabolism results from a consortium of two symbionts that are otherwise free-living.

Many of the foraminiferans and radiolarians possess symbiotic algae. In some foraminiferans and radiolarians several different symbiotic species of algae may live within the protozoan cytoplasm. During the day the endosymbionts are distributed in the pseudopodial network, but at night they are withdrawn close to the main body of the cell or into the shell. Many thousands of these algae may exist within a single protozoan, and a significant amount of the products of photosynthesis (e.g., glucose, alanine, maltose) are transferred from the algae to the protozoan. Indeed, in some circumstances, the protozoan can survive on this source of energy if deprived of food, although its growth may be impaired.

Mixotrophy in Planktonic Protozoans

Some mixotrophs (e.g., the planktonic protozoans Dinobryon and Ochromonas) also feed on bacteria but are phototrophs first and foremost. Thus, they are "true" mixotrophs, unlike protozoans that exploit the photosynthetic capability of symbiotic organisms and cannot carry out this form of metabolism themselves. Many planktonic marine and freshwater mixotrophs feed voraciously on bacteria. In some lakes these protists may be the main consumers of bacteria suspended in surface waters. It is believed that this ingestion of bacteria provides the mixotrophs not only with an additional source of carbon to supplement what is gained by photosynthesis but also with phosphorus and nitrogen, which are often scarce in biologically productive waters, and possibly with vitamins, which are essential to photosynthesis. Bacteria are more efficient at taking up these nutrients because they have a higher surface-to-volume ratio than protists. Thus, one way for the protists to acquire essential nutrients is to consume the bacteria.

Reproduction and Life Cycles

Asexual reproduction is the most common means of replication by protozoans. The ability to undergo a sexual phase is confined to the ciliates, the apicomplexans, and restricted taxa among the flagellated and amoeboid organisms. Moreover, sexual reproduction does not always result in an immediate increase in cell numbers but may simply be a means of exchanging genetic material between individuals of the same species (i.e., conjugation). Free-living protozoans normally resort to sexual reproduction only when environmental conditions become adverse, because this mode of reproduction enhances genetic variation through mechanisms such as mutation and chromosomal crossing over. These processes maintain genetic diversity within a population, which supports population fitness and survival. When food and other conditions are favourable, asexual reproduction occurs.

Mechanisms of Asexual Reproduction

Asexual reproduction in free-living species usually involves nuclear division and the division of the cell into two identical daughter cells of equal size by binary fission. In parasitic protozoans and some free-living species, multiple fission, resulting in the production of many offspring that may not resemble the parent cell, is normal. During the cycle of growth and division, the protozoan undergoes a series of identifiable phases: a division phase, a growth phase during which the cell increases substantially in size, a phase of DNA synthesis, and a phase of preparation for division, which extends from the end of DNA synthesis until the initiation of division. The division of the cytoplasm (cytokinesis) is preceded by the division of the nucleus or nuclei.

The plane of division in protozoan cells varies among the different groups and is of taxonomic significance. The ciliates normally divide in an equatorial, or transverse, plane, thereby maintaining the correct number of ciliary rows, or kineties. The cell mouth and any specialized cilia around it are replicated in different ways among the various ciliate groups, depending on the complexity of the cytostome. The replication of the cytostome precedes the division of the cytoplasm. Some ciliates (e.g., Colpoda) divide within thin-walled reproductive cysts into two daughter ciliates, each of which then divides so that the cyst contains four progeny, which are released when the cyst wall ruptures.

The sedentary suctorians do not reproduce by binary fission, because the production of an identical nonswimming offspring would rapidly lead to overcrowding. They instead produce single ciliated offspring, called swarmers, by a process called budding. Budding can occur endogenously, in which the bud forms within the parent and is ejected when mature, or exogenously, in which the swarmer is formed outside the parent. The swarmers swim away from the parent, settle on a substrate, lose their cilia, and develop feeding tentacles and an attaching stalk.

Naked amoebae have no fixed plane of division but simply round up and divide into two basically equal halves. The testate amoebae, which live in single-chambered shells, or tests, exude the daughter from the aperture of the shell. In species that have a shell formed from silica plates, the daughter contains the plates used to produce the shell but remains attached to the mother cell until the shell is fully formed, when the final severing of the cytoplasm between the individuals occurs. Some of the testate amoebae live inside proteinaceous shells. There too, the new shell is secreted before binary fission is completed.

The foraminiferan and radiolarian amoebae have evolved multiple fission. Both produce many flagellated swarmers, or zoospores. The common planktonic foraminiferan Globigerinoides sacculifer, for example, can produce 30,000 swarmers at one time. Each swarmer is about 5 micrometres (0.005 mm) long. In planktonic species the parent usually loses buoyancy and sinks by shedding spines and withdrawing the complicated pseudopodial network into the shell. The swarmers are produced in deep water and migrate upward as they mature. Each secretes a shell around itself, which is added to as the organism grows.

Mechanisms of Sexual Reproduction

The foraminiferans are unusual among free-living protozoans in that a sexual phase is a regular part of the life cycle, alternating with an asexual phase. During the life cycle two types of swarmers are produced. One type, zoospores, have half the number of chromosomes of the parent (i.e., they are haploid); they grow until they become mature adults and can produce and release large numbers of gametic swarmers. These gametes are identical (isogamous) but are functionally comparable to the eggs and sperm of higher organisms. The gametic swarmers fuse in pairs, thus restoring the full complement of chromosomes (i.e., they are diploid), and each individual grows, matures, and ultimately produces haploid zoospores.

Sexual reproduction among other protozoans is not widespread and can involve identical gametes (isogamy) or distinct male and female gametes (anisogamy or heterogamy). The female gametes are usually larger and are stationary, whereas the male gametes are smaller, produced in larger numbers, and motile.

Conjugation in Ciliates

Sexual reproduction among the ciliated protozoans takes the form of conjugation. The process does not result in an increase in numbers but is a simple exchange of genetic material between two individual cells. Conjugation occurs only between compatible mating strains within a species, and each species may contain many mating strains. Before conjugation occurs, special chemical signals, called gamones, are released by some ciliates. The gamones cause compatible mating strains to undergo processes that facilitate conjugation. In other ciliates, such as Paramecium, gamones are bound to the cell surface and elicit their responses when the ciliates make physical contact.

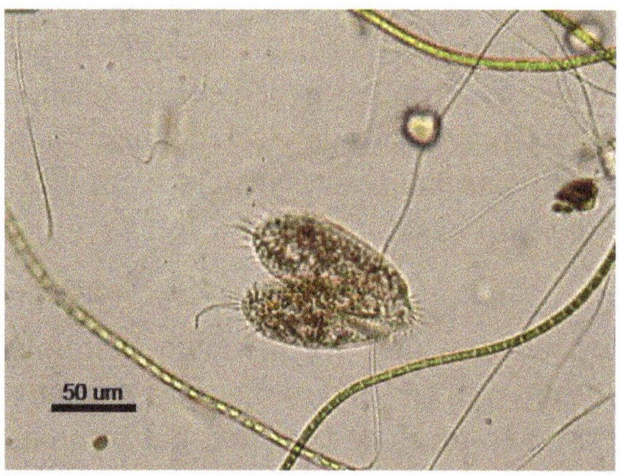

Two hypotrichs in conjugation.

During conjugation, two ciliates line up side by side. The macronucleus, which plays no part in the process, disintegrates. A series of nuclear divisions of the micronuclei in each ciliate then ensues, including meiosis, during which a number of haploid micronuclei are produced in both cells. All but one of these haploid micronuclei disintegrate. The remaining haploid micronucleus in each cell then divides through mitosis into two haploid nuclei (gamete nuclei). A bridge of cytoplasm forms between the two ciliates, and one gametic nucleus from each cell passes into the other cell. The two gametic nuclei in each cell unite, thus restoring the diploid number of chromosomes. The micronucleus undergoes two mitotic divisions to produce four micronuclei: two of these will form the new micronuclei of the cell, and two are destined to become the macronucleus. Following the process of conjugation, normal binary fission proceeds. The number of macronuclei and micronuclei formed is dependent on the species and remains the same as the original number.

Autogamy and Modified Conjugation

When no suitable mating partner is available, ciliates may undergo a form of conjugation called autogamy, in which all the nuclear processes described above occur. But, because only one individual is involved, there is no exchange of gametic nuclei. Instead, the two gametic nuclei within the cell unite to form the restored micronucleus.

Specialized sedentary suctorian ciliates practice a modified form of conjugation. The conjugating individuals differ in appearance. The macroconjugants resemble the normal feeding individuals, and the microconjugants resemble the swarmers, although smaller. When a microconjugant locates a macroconjugant, it enters and fuses with it. This is quite different from the temporary association between two cells that occurs during sexual reproduction in most ciliates.

Parasitic protozoan Life Cycles

As is common with other parasitic organisms, parasitic protozoans face the problem of how to disperse from one host to another. In order to increase the probability of finding more hosts, most parasitic protozoans reproduce in high numbers. A representative life cycle of a parasitic protozoan can be found in members of the parasitic group Apicomplexa. These protozoans have a complex life cycle that involves a series of stages characterized by episodes of asexual multiple division called schizogony. In the parasite Plasmodium, for example, this phase of the life cycle occurs in the liver and red blood cells of humans. The parasite (sporozoite) enters the host's cells and grows while feeding on the cell contents. It then undergoes a multiple asexual division (schizogony) into many individuals (merozoites). The host's cell wall ruptures, permitting each individual to invade a new red blood cell and repeat the process.

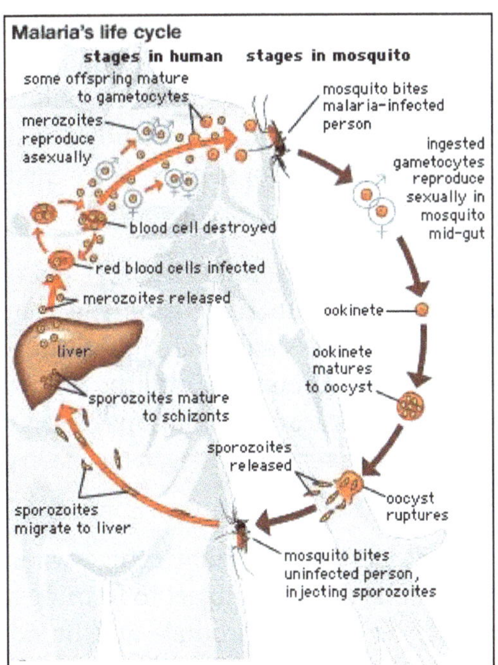

Life cycle of a malaria parasite.

In certain merozoites a sexual cycle is eventually initiated inside the red blood cell, and male and female gametes are produced. The male gametes (microgametocytes) are small, while the female gametes (macrogametocytes) are larger. The life cycle continues if the gametocytes are taken up by a feeding female mosquito of the genus Anopheles. Only the gametocytes can infect the mosquito. Inside the mosquito's gut the haploid gametes fuse to form a diploid zygote, which then undergoes sporogony, a process of multiple divisions in which many sporozoites are produced. The sporozoites migrate to the salivary glands of the insect and are injected into a new host when the mosquito feeds again. They are carried by the blood to the liver, where they undergo their first schizogony inside liver cells, thereafter invading the red blood cells for repeated cycles of schizogony.

Flagellated protozoan parasites reproduce almost exclusively by asexual means and do not appear to have a sexual phase in their life cycles. There is, however, evidence of genetic exchange between certain subspecies of Trypanosoma brucei.

Adaptations

For the most part, parasitic protozoans live in a fairly constant environment. Temperature fluctuates very little, or not at all, inside the host, desiccation is not a risk, and food is in constant supply. Free-living protists, on the other hand, face short- or long-term changes in temperature, aquatic acidity, food supply, moisture, and light. Many protozoans respond to adverse environmental conditions by encysting: they secrete a thick, tough wall around themselves and effectively enter a quiescent state comparable to hibernation. The ability to form a resistant cyst is widespread among diverse protistan groups and probably developed early in their evolutionary history. Resting cysts also are easily carried by the wind and form an important means of dispersal for species that live in the soil or are common in ephemeral ponds and pools. In climates with distinct cold seasons, the cyst may be an important phase in the annual life cycle.

The cyst wall is composed of a varying number of layers, the components of which are dependent on the species. During the encystment process, the protozoan cell undergoes a series of changes that considerably reduce the complexity of the organism. Flagellated organisms and ciliates lose their flagella and cilia, the contractile vacuole and food vacuoles disappear, and the distribution of organelles within the cell may be reorganized. In some species the cell volume reduces considerably. These changes are reversed during the process of excystment.

Certain marine planktonic tintinnids are programmed to break out of their cysts en masse at times of the year when the food supply is abundant. Helicostomella subulata, for example, excysts in June in temperate waters and becomes numerous from July through October. It encysts again in October, sinking to the sediments, where it remains until the following year. The cyst is a normal part of the annual life cycle, and even laboratory populations of this ciliate encyst at the same time as the natural population. This type of life strategy pattern has been demonstrated in several other ciliates and in some amoebae.

For soil-dwelling protozoans the cyst is an important refuge when soil moisture disappears or when soil water becomes frozen. In soils that are subject to freezing and periodic short-term thawing, the protozoans rapidly excyst, feed, and reproduce and then encyst again when soil water becomes temporarily unavailable to them.

The cyst plays an important role in the life cycles of several parasitic protozoans that have a free-living dispersal stage, such as Entamoeba histolytica and Cryptosporidium. The cysts are excreted in the host's feces and survive in water or soil. Humans are usually infected through drinking contaminated water or eating raw fruit and vegetables grown where human feces are used as fertilizer.

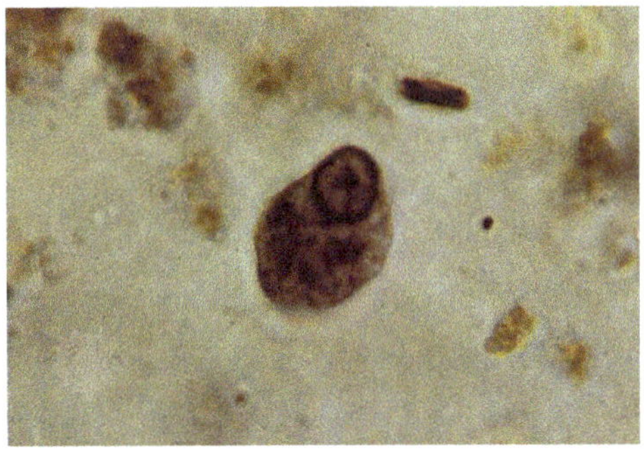

Entamoeba coli.

Some freshwater protozoans, especially the ciliates Spirostomum, Loxodes, and Plagiopyla, avoid unpleasant conditions, especially lack of oxygen, by abandoning their bottom-dwelling way of life and swimming upward to position themselves at a level where some oxygen is available but where they are not in direct competition with planktonic species. They remain there until oxygen again becomes available on the lake bottom, at which time they migrate downward.

The widespread occurrence of mixotrophy involving algal symbiosis and the retention and sequestration of the plastids of photosynthetic prey by planktonic protozoans is believed to be an adaptation to waters where food is limited. Ciliates that retain plastids appear to be far more common in waters where food is scarce than in productive waters. An inverse relationship exists between this form of mixotrophy and the productivity of the ecosystem.

Evolution and Paleontology

Protists were a dominant form of life on Earth 1.5 billion years ago. While protozoans evolved early and have survived to the present day as unicellular organisms, they have undoubtedly undergone considerable evolutionary change. That many species must have become extinct as others appeared can be deduced from the limited fossil record of protozoans. Extinct fossil foraminiferan species, for example, number around 34,000, while there are only about 4,000 described living species.

Only a small number of protozoans, most of which are testate amoebae, have left fossil remains. The calcareous shells of the foraminiferans and calcium-secreting coccolithophores (a group of algae), for example, produced substantial geologic strata in the chalk formed during the Cretaceous Period (145.5 million to 65.5 million years ago) and the well-developed foram-limestones of the Paleozoic Era (542 million to 251 million years ago), Early Cretaceous Epoch (145.5 million to 99.6 million years ago), and Cenozoic Era (65.5 million years ago to the present). The fossil-forming radiolarians date to late Precambrian times, and the testate lobose amoeba Melanocyrillium dates to the late Precambrian geologic record of the Grand Canyon in northwestern Arizona, U.S. The testate amoeba Nebela is found in deposits from the Cretaceous Period.

The most abundant and important fossil protozoans are the foraminiferans. This entirely marine group is extremely important as stratigraphic markers in oil exploration. Because species have appeared and then become extinct frequently during geologic history and because they have fairly wide geographic distribution, particularly planktonic species, their value is in showing distinct phases in geologic history and, with specific species, in typifying particular beds of rock or strata. Foraminiferans are also important in the reconstruction of paleoceanographic circulation patterns.

The poor fossil record of protozoans has hampered attempts at unraveling the complexities of their evolution. Modern biochemical and electron microscopy techniques, however, are providing evidence for new affinities between groups and are elucidating possible evolutionary pathways. Comparisons of flagellar structures, mitochondria, and nuclear and plastid characteristics in conjunction with ribosomal RNA (ribonucleic acid) sequences are revealing the relationships of various taxa.

The ancestral eukaryote organism is thought to have been an amoeboid creature that relied on anaerobic or microaerophilic metabolism (microaerophilic organisms survive on only very small amounts of oxygen). The evolution of mitochondria (the centres of aerobic respiration in the cell) as organelles from endosymbiotic bacteria and the establishment of oxidative pathways allowed a more efficient cellular energy balance, which led the way to the evolution of an enormously diverse array of eukaryotic organisms. Some of the early amoeboid eukaryotes developed flagella to enhance their food-gathering abilities and to provide a more efficient mode of propulsion. The flagellates gradually evolved different ways of life, and their structures became modified accordingly. As phagotrophs that ingested bacteria for food, they in some cases came to establish symbiotic associations with photosynthetic species, and ultimately the endosymbionts became plastids within the cell. Some of the flagellates came to depend entirely on photosynthesis and to abandon heterotrophy completely, though many still retain both heterotrophic and autotrophic nutrition as mixotrophs. (Some present-day mixotrophs, however, may be only secondarily mixotrophic, having reestablished heterotrophy in conjunction with photosynthesis).

A considerable number of protozoans became parasitic, a mode of life that evolved independently among the protozoans many times. Ciliates and amoebae became symbionts in the intestinal tracts of both vertebrates and invertebrates as a result of surviving the digestive enzymes of the predator. (Most present-day parasites among these protists are intestinal parasites.) Once inside the intestine of the host, they multiplied and gradually, through mutation and selection, came to rely on the resistant cyst as a means of survival and dispersal, losing the ability to survive in a free-living feeding form.

The process of parasitism probably arose in several independent cases. The trypanosomes, for example, evolved from free-living forms, adapting to life in the alimentary canal of primitive invertebrates during late Precambrian times (570 million years ago). They evolved with their hosts, becoming symbionts in a wide variety of invertebrates, including annelids, nematodes, and mollusks. It was in the insects, however, that they underwent their most extensive evolutionary explosion into two groups. At this stage they were transmitted from insect to insect by resistant cysts passed in the feces and ingested by subsequent hosts. When insects developed the habit of sucking vertebrate blood, which is believed to have occurred about 40 million years ago, the protozoan symbionts that lived in the gut entered the blood of vertebrates, probably as feces left by the insect were rubbed into the wound. The blood provided a rich environment for the flagellates and thus evolved the two-host life cycles seen today in the Leishmania and Trypanosoma groups.

The apicomplexans, which also inhabit the blood of vertebrates at some stage in their life cycle, probably evolved from a basal primitive stock seen today as the gregarines, which are parasites of invertebrates. They gave rise to a group of parasitic organisms of which the coccidia, with a one-host life cycle, are primitive survivors. At first these protozoans lived in the gut of their vertebrate host, but they gradually began invading host tissues and eventually became adapted to spending part of their life cycle in the bloodstream. There they were taken up by blood-feeding insects, and an insect vector host became incorporated into the life cycle. Associated modifications in the reproductive pattern, as seen in Plasmodium, which belongs to the Haemosporina, also occurred. This series of events appears to have happened at least twice in the evolution of apicomplexan life cycles.

Classification

General Principles

A fundamental shift in protozoan taxonomy occurred in 1990, when American microbiologist Carl Woese and colleagues revolutionized the world of biology with the three-domain classification system of life. Based on sequences of rRNA (ribosomal RNA), molecules present in all organisms as part of the protein-manufacturing machinery, Woese's classification system revealed three major evolutionary groups of life on Earth, one of which is eukaryotic (the Eukarya) and two of which are prokaryotic (Eubacteria and Archaea). It is generally accepted that the Eubacteria (now Bacteria) are the most distant genetic group of the three. The three-domain system has largely replaced the previous five-kingdom system of American biologist Robert H. Whittaker, which is based on morphology (e.g., mode of nutrition) rather than phylogeny (the history of the evolution of a species or group).

Classification within the domain Eukarya also experienced a shift from morphology-based approaches to an emphasis on phylogenetic relationships. On the basis of that advance, the former eukaryotic kingdoms Animalia, Plantae, and Fungi are no longer separated from protists. Instead, each of those major groups of multicellular life is classified within a supergroup of Eukarya along with a protistan group. For example, plants are classified within the supergroup Archaeplastida, alongside some examples of unicellular algae, and the animals and fungi are classified within the supergroup Opisthokonta, along with single-celled choanoflagellates.

Current scientific approaches have produced a nested yet nonhierarchical picture of biological classification, in stark contrast to the heirarchical scheme of Linnean taxonomy, which specifies somewhat arbitrary universal ranks of classification (e.g., Kingdom, Phylum, Class, Order). This development has caused many biologists to abandon the Linnean system, primarily at the higher levels of classification, rather than at the genus and species levels. The validity and utility of the Linnean taxonomic approach continues to be a source of debate among biologists. Major developments in the classification of protists in the 1990s and early 2000s did not use hierarchical schemes.

Protistan systematics remains a subject of debate and change. Protists comprise a large and sometimes unwieldy assemblage, and assignments of species to particular taxa change as new genomic approaches, biochemical techniques, and electron microscopy studies provide more details on the affinities of various species.

Diagnostic Features

A general consensus regarding the classification of eukaryotes (with emphasis on protists) was published by the International Society of Protistologists (ISOP) in 2005. This classification system of eukaryotic taxa divides species into monophyletic groups, or clades. Monophyletic groups contain a common ancestor and all its descendants. This type of grouping is in contrast to paraphyletic groupings (consisting of a common ancestor and some of its descendants) and polyphyletic groupings (consisting of taxa that do not share a common ancestor). Monophyletic groups are defined by the possession of shared, derived traits known as apomorphies.

The classification scheme introduced by the ISOP defines six monophyletic supergroups of eukaryotes: Archaeplastida, Excavata, Chromalveolata, Amoebozoa, Rhizaria, and Opisthokonta. Using this scheme, the protozoans and algae are clearly polyphyletic. Former groupings defined by previous classification schemes are no longer recognized in the 2005 system, such as the former phylum Sarcomastigophora, which grouped paraphyletic lineages of photosynthetic and nonphotosynthetic taxa.

Balantidium Coli

Balantidium coli is a parasitic species of ciliate alveolates that causes the disease balantidiasis. It is the only member of the ciliate phylum known to be pathogenic to humans.

Morphology

Balantidium coli has two developmental stages, a trophozoite stage and a cyst stage. In trophozoites, the two nuclei are visible. The macronucleus is long and sausage-shaped, and the spherical micronucleus is nested next to it, often hidden by the macronucleus. The opening, known as the peristome, at the pointed anterior end leads to the cytostome, or the mouth. Cysts are smaller than trophozoites and are round and have a tough, heavy cyst wall made of one or two layers. Usually only the macronucleus and sometimes cilia and contractile vacuoles are visible in the cyst. Living trophozoites and cysts are yellowish or greenish in color.

Transmission

Balantidium is the only ciliated protozoan known to infect humans. Balantidiasis is a zoonotic disease and is acquired by humans via the feco-oral route from the normal host, the pig, where it is asymptomatic. Contaminated water is the most common mechanism of transmission.

Role in Disease

Balantidium coli lives in the cecum and colon of humans, pigs, rats and other mammals. It is not readily transmissible from one species of host to another because it requires a period of time to adjust to the symbiotic flora of the new host. Once it has adapted to a host species, the protozoan can become a serious pathogen, especially in humans. Trophozoites multiply and encyst due to the dehydration of feces.

Infection occurs when the cysts are ingested, usually through contaminated food or water. Balantidium infection in immunocompetent individuals is not unheard of, but it rarely causes serious disease of the gastrointestinal tract. It can thrive in the gastrointestinal tract as long as there is a balance between the protozoan and the host without causing dysenteric symptoms. Infection most likely occurs in people with malnutrition due to the low stomach acidity or people with compromised immune systems.

In acute disease, explosive diarrhea may occur as often as every twenty minutes. Perforation of the colon may also occur in acute infections which can lead to life-threatening situations.

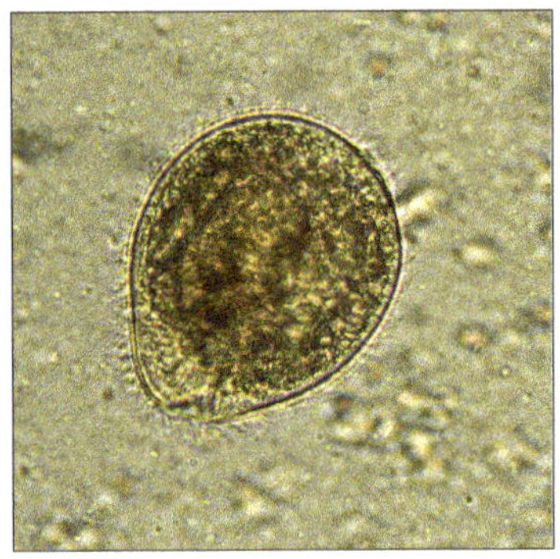

Balantidium coli as seen in a wet mount of a stool specimen. The organism is surrounded by cilia.

Life Cycle

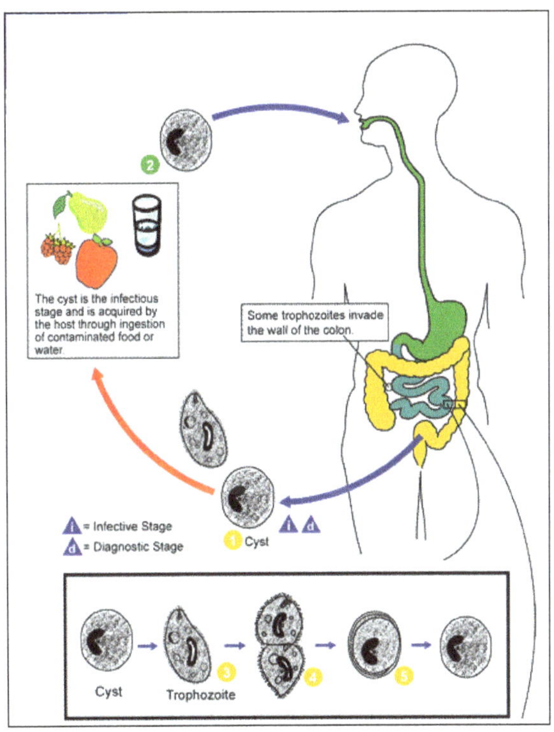

Balantidium coli life cycle.

Infection occurs when a host ingests a cyst, which usually happens during the consumption of contaminated water or food. Once the cyst is ingested, it passes through the host's digestive system. While the cyst receives some protection from degradation by the acidic environment of the stomach through the use of its outer wall, it is likely to be destroyed at a pH lower than 5, allowing it to survive easier in the stomachs of malnourished individuals who have less stomach acid. Once the cyst reaches the small intestine, trophozoites are produced. The trophozoites then colonize the large intestine, where they live in the lumen and feed on the intestinal flora. Some trophozoites invade the wall of the colon using proteolytic enzymes and multiply, and some of them return to the lumen. In the lumen trophozoites may disintegrate or undergo encystation. Encystation is triggered by dehydration of the intestinal contents and usually occurs in the distal large intestine, but may also occur outside of the host in feces. Now in its mature cyst form, cysts are released into the environment where they can go on to infect a new host.

Epidemiology

Balantidiasis in humans is common in the Philippines, but it can be found anywhere in the world, especially among those that are in close contact with swine. The disease is considered to be rare and occurs in less than 1% of the human population. The disease poses a problem mostly in developing countries, where water sources may be contaminated with swine or human feces.

Leishmania

Leishmania is any of several species of flagellate protists belonging to the genus Leishmania in the order Kinetoplastida. These protists are parasites of vertebrates, to which they are transmitted by species of Phlebotomus, a genus of bloodsucking sand flies. The leishmanial parasites assume two forms: a round or oval leishmanial stage, which lives and multiplies in the vertebrate host; and an elongate, motile, flagellated organism called a leptomonad, which is found in the alimentary tract of the sand fly. In their leishmanial stage, the organisms are taken in with the meal of the fly, and they develop into leptomonads in the fly's stomach and multiply there. They eventually migrate to the fly's mouthparts, from which the leptomonads enter the wound made at the next feeding, thus initiating a new infection.

There are three separate species in the genus Leishmania: these three species look quite alike but cause three different human diseases that are collectively called leishmaniasis. L. donovani, which attacks the liver, spleen, bone marrow, and other viscera, causes kala-azar (q.v.) in Africa, Europe, and Asia. L. tropica causes oriental sore (q.v.) in Africa, Europe, and the East; lesions that range from pimples to large ulcers are formed on the skin of the hands, feet, legs, and face. L. brasiliensis, the cause of American leishmaniasis in Central and South America, produces similar skin lesions but also causes deeper lesions of the oral and nasal mucous membranes.

Leishmania Donovani

Leishmania donovani is a species of intracellular parasites belonging to the genus Leishmania, a group of haemoflagellate kinetoplastids that cause the disease leishmaniasis. It is a human blood parasite responsible for visceral leishmaniasis or kala-azar, the most severe form of leishmaniasis. It infects the mononuclear phagocyte system including spleen, liver and bone marrow. Infection is transmitted by species of sandfly belonging to the genus Phlebotomus in Old World and Lutzomyia in New World. Therefore, the parasite is prevalent throughout tropical and temperate regions including Africa (mostly in Sudan), China, India, Nepal, southern Europe, Russia and South America. It is responsible for thousands of deaths every year and has spread to 88 countries, with 350 million people at constant risk of infection and 0.5 million new cases in a year.

L. donovani was independently discovered by two British medical officers William Boog Leishman in Netley, England, and Charles Donovan in Madras, India, in 1903. However, the correct taxonomy was provided by Ronald Ross. The parasite requires two different hosts for a complete life cycle, humans as the definitive host and sandflies as the intermediate host. In some parts of the world other mammals, especially canines, act as reservoir hosts. In human cell they exist as small, spherical and unflagellated amastigote form; while they are elongated with flagellum as promastigote form in sandflies. Unlike other parasitic protists they are unable to directly penetrate the host cell, and are dependent upon phagocytosis. The whole genome sequence of L. donovani obtained from southeastern Nepal was published in 2011.

Discovery

One of the earliest known epidemics of L. donovani infection (kala-azar as it was called in Hindi) was known in India just after the Indian Rebellion of 1857. The first medical record was however only in 1870 by British medical officers from Assam. In 1900, an English soldier stationed at Dum Dum, West Bengal, died at the Army Medical School in Netley, England. The autopsy was performed by William Boog Leishman. He processed the tissue sample of the enlarged spleen using a staining technique (now known as Leishman's stain) which he had just developed, and discovered the protozoan parasites using microscopy. But he mistakenly considered the parasites to be degenerate trypanosomes, already known protozoan parasites in Africa and South America. In 1903, Leishman published his discovery of "trypanosomes in India" in the British Medical Journal, which appeared on 11 May. Another British medical officer Charles Donovan, who was serving in the Indian Medical Service, had found the parasites in April of that year at the Government General Hospital in Madras. After reading Leishman paper, Donovan confirmed on 17 June that the parasites (by then known as "Leishman bodies") were definitely the causative agents of kala-azar. He wrote a commentary of his discovery in relation to that of Leishman in the same journal, that appeared on 11 July 1903. Soon a controversy arose as to whom such a monumental discovery should be credited. Donovan sent some of his slides to Ronald Ross, who was at the time in Liverpool, and to Alphonse Laveran at the Pasteur Institute in Paris. Laveran and his colleague Félix Mesnil identified the protozoan (and yet wrongly) to be members of Piroplasmida, and gave the scientific name Piroplasma donovanii. It was Ross who resolved the conflict of priority in the discovery and correctly identified the species as member of the novel genus Leishmania. He gave the popular name "Leishman-Donovan bodies", and subsequently the valid binomial Leishmania donovani, thereby equally crediting the two rivals.

Structure

Leishmania donovani is a unicellular eukaryote having a well-defined nucleus and other cell organelles including a kinetoplast and a flagellum. Depending on its host it exists in two structural variants, as follows:

1. Amastigote form found in the mononuclear phagocyte and circulatory systems of humans. It is an intracellular and non-motile form, being devoid of external flagellum. The short

flagellum is embedded in the anterior end without projecting out. It is oval in shape, and measures 3–6 μm in length and 1–3 μm in breadth. The kinetoplast and basal body lie towards the anterior end.

2. Promastigote is formed in the alimentary tract of the sandfly. It is an extracellular and motile form. It is considerably larger and more highly elongated, measuring 15–30 μm in length and 5 μm in width. It is spindle-shaped, tapering at both ends. A long flagellum (about the body length) is projected externally at the anterior end. The nucleus lies at the centre, and in front of which are kinetoplast and basal body.

Infection and Life Cycle

Leishmania donovani is a digenetic parasite passing its life cycle in two different hosts.

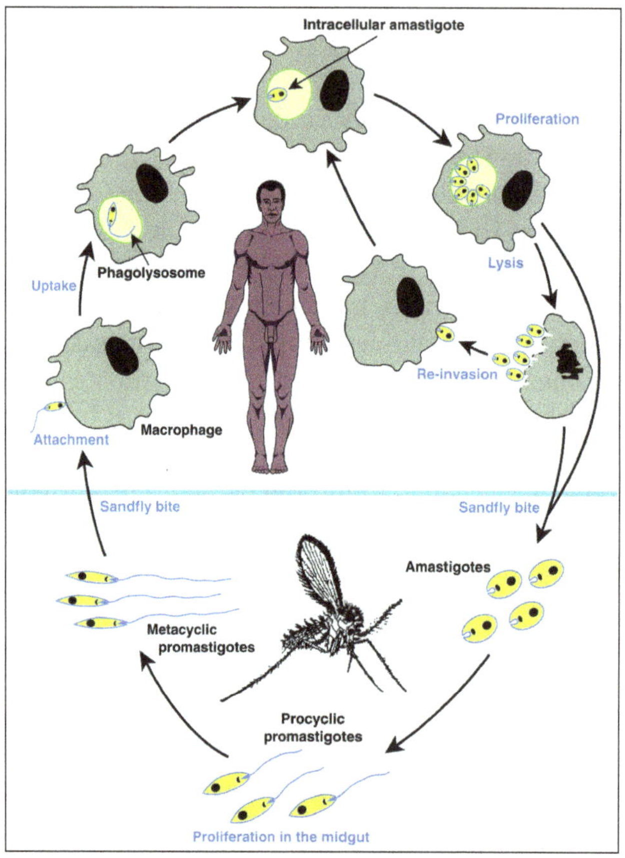

Life cycle of Leishmania donovani.

Definitive Host

In humans the metacyclic promastigotes are injected by sandfly through the skin during its blood meal. When sandfly bites using its proboscis it ejects the parasites that are stored inside the hollow tube. Some promastigotes may enter the blood stream directly where some are destroyed by macrophagic cytolysis. But many are also taken up through phagocytosis by mononuclear phagocytes in liver, spleen and bone marrow. Inside the cells they undergo spontaneous transformation into oval-shaped amastigotes. Granulocytes selectively kill the promastigotes by oxidative mechanism, while amastigotes are resistant. Then the surviving amastigotes undergo cell division using simple binary fission. Multiplication continues until the host cell can no longer hold and ruptures. In a fully congested cell there can be as many as 50 to 200 amastigotes, which are released into tissue cavities. Each individual amastigote is then capable of invading fresh cells. As a result, the entire tissue is progressively infected and destroyed. A number of free amastigotes then enters the blood stream where many are phagocytosed by macrophages. These free and phagocytosed amastigotes in peripheral blood are then sucked up by blood-feeding sandfly.

Intermediate Host

L. donovani undergo further development only in the digestive tract of the female sandfly. Hence only females are responsible for transmitting the infection. Once the amastigotes are ingested, they enter the midgut of the sandfly. Then they undergo structural modification into flagellated promastigotes, becoming larger and considerably elongated. They get attached to the gut epithelial lining where they multiply rapidly by binary fission. (They are also capable of sexual reproduction by genetic hybridisation in the sandfly gut.) They then migrate back towards the anterior part of the digestive system such as pharynx and buccal cavity. This process is known as anterior station development, which is unique in Leishmania. A heavy infection of pharynx can be observed within 6 to 9 days after initial blood meal. The promastigotes become infective only by this time, and the event is called the metacyclic stage. The metacyclic promastigotes then enter the hollow proboscis where they accumulate and completely block the food passage. Immediately upon biting a human, the parasites are released, which invariably results in infection. The stages of development in sandfly can be described as follows:

1. Soon after entering the gut, the amastigotes get coated with peritrophic matrix, which is composed of chitin and protein complex. This protects the parasites from the digestive enzymes of the host.

2. The amastigotes travel as far as the abdominal midgut and first transform into a weakly motile "procyclic promastigotes" on the gut wall within 1–3 days.

3. The young promastigotes secrete a neuropeptide that stop peristalsis of the gut. The surface lipophosphoglycan (LPG) of the promastigote serves as an attachment to the gut epithelium. These factors prevent the expulsion of promastigotes during excretion of the insect.

4. During 4–7 days the peritrophic matrix is degraded by the activity of chitinases. This release the more actively motile "nectomonad promastigotes" which migrate anteriorly until they reach the opening of the thoracic gut.

5. Another transformation takes place by which they turn into "leptomonad promastigotes". These are fully motile and capable of binary fission. Multiplication and migration towards thoracic midgut cause congestion of the pharynx and buccal cavity. Here they secrete promastigote secretory gel (PSG), which is composed of soluble acid phosphatase and phosphoglycoprotein.

6. After 6–9 days the promastigotes become metacyclic. Some are also transformed into non-replicating promastigotes, which also become metacyclic. The sandfly is able to regurgitate and eject the parasites from its proboscis with the help of PSG when it bites.

Reservoir Host

Dogs are known to be susceptible to L. donovani infection. Especially in the New World, infection is a zoonotic disease, involving different canine species, including domestic dog and the two fox species, Lycalopex vetulus and Cerdocyon thous. In the Mediterranean region domestic dogs and the three fox species Vulpes vulpes, V. corsac and V. zerda are common reservoir hosts. In Africa and Brazil, some marsupials and rodents are also reported to harbour L. donovani.

Epidemiology

It is estimated that visceral leishmaniasis (VL) affects more than 100 million people worldwide, with 500,000 new cases and more than 50,000 deaths each year. Although L. donovani is only the second-most prevalent Leishmania causing VL, it is the most dangerous form and directly fatal to humans. Over 90% of reported cases are from India, Bangladesh, Nepal, Sudan and Brazil. In India it is prevalent in the eastern region including Bihar, West Bengal, eastern Uttar Pradesh, Assam and foothills of Sikkim. It is responsible for tens of thousands of mortality among Africans in eastern and southern parts of Sudan. During the epidemic of 1984–1994 death toll was as high as 70% in the Sudanese population. Moreover, due to emergence of drug resistance the prevalence is not subsiding, and in fact has spread to central Europe. For example, during the late 1990s hundreds of cases were reported in Switzerland.

Pathogenicity

L. donovani is the causative agent of visceral leishmaniasis, traditionally known as kala-azar ("black fever", particularly in India), because of its characteristic symptoms. The disease is highly lethal if not treated properly. The incubation period generally ranges from 3 to 6 months, and in some cases may be over a year. In Indian leishmaniasis, incubation can be as short as 10 days. The target cells are those of mononuclear phagocyte system. The two main tissues of infection are spleen and liver. Clinical symptoms include pyrexia (recurring high fever which may be continuous or remittent), enlargement of spleen and liver, and heavy skin pigmentation which darkens the physical appearance (the reason for naming "black fever"). To a lesser extent mucosa of the small intestine and lymph nodes are also invaded by the parasite. Morphological symptoms are noticeable particularly on facial and abdominal regions. Skin becomes coarse and hard. In African infections, warty eruptions are common. In a fully developed stage, the patient shows emaciation and anaemia. Where medical facilities are poor, mortality can be as high as 75–95% within 2 years of epidemics. The disease is often accompanied by complications with dysentery, tuberculosis, septicaemia and even HIV infection.

Cellular Invasion and Immunological Response

Amastigotes of L. donovani enter macrophages via a Rac1- and Arf6-dependent process, and are found in phagocytic vacuoles that interact with endosomes and lysosomes and acquire lysosomal features. During phagocytosis by macrophages, the promastigotes inhibit the formation of the phagolysosome, a cellular product by which invading pathogens are removed. The promastigote can do this using its glycolipid lipophosphoglycan (LPG) on its cell membrane. LPG causes disorganisation of F-actin and disruption of phagosomal lipid microdomains. They are capable of evading the microbicidal actions of macrophages, which can kill ordinary pathogens using reactive nitrogen and oxygen intermediates. They effectively subvert the production of reactive oxygen species. In this way the amastigotes are able to survive and replicate inside these primary immune systems. The parasites manipulate the cell signalling pathway of the macrophages, such as down-regulating of Jak/stat signalling, NO and TNF-α production, and also by blocking the NF-κB-dependent pathway. There are two major mechanisms of immune evasion such as induction of immune suppressive IL-10 responses and the generation of poor and functionally impaired CD8(+) T-cell responses.

Treatment

The conventional treatment method is an intravenous injection with antimony compounds, such as pentostam. Unfortunately, these chemotherapeutics are so poisonous that about 15% of the patients die from the treaments. To compound the situation, drug resistance has evolved in the parasites against the traditional antimonials. According to rough estimates, about 40% of patients in India are already resistant to this therapy. Another antimicrobial drug amphotericin B is also commonly used. Liposomal amphotericin B (L-AmB) has been a drug of choice in India, but is practically useless in Africa because of low effectiveness in the African strain of the parasite. Further, amphotericin B has severe adverse effects. Its acute effects includes nausea, vomiting, rigors, fever, hypertension or hypotension, and hypoxia, and its chronic effect is nephrotoxicity. In 1999 an anticancer drug miltefosine was demonstrated to be highly effective (95% cure rate) among Indian patients. This was the first time an oral drug is effective for visceral leishmaniasis. Clinical trials showed that the new drug is relatively harmless. The most adverse effects were only vomiting and diarrhoea in 20–28% patients, which were rather mild. The drug has been officially approved in India. The recommended dosage is 100 mg per day over a period of four weeks.

Evolution

L. donovani is now considered to be a complex species as indicated by different pathological symptoms occurring in different geographical areas where the species of the vector sandfly are also different. However, none of the parasites are morphologically distinguishable, except by molecular analysis. Molecular data show that genotype is strongly correlated with geographical origin. DNA sequencing of different geographical strains indicates that the protozoan complex can be classified into two valid taxons, L. donovani and L. infantum. The genus Leishmania most likely originated in South America, from where it migrated to Asia. L. donovani and L. infantum diverged ~1 Mya, with further divergence of infraspecific genetic groups between 0.4 and 0.8 Mya.

Leishmania Mexicana

Leishmania mexicana belongs to Leishmania genus and is the causal agent of cutaneous leishmaniasis in Mexico and central America.

Leishmania mexicana is an obligate intracellular protozoan parasite that causes the cutaneous form of leishmaniasis. This species of Leishmania is found in America. The infection with L. mexicana occurs when an individual is bitten by an infected sandfly that injects infective promastigotes, which are carried in the salivary glands and expulsed by the proboscis, directly to the skin.

The life cycle of this and other Leishmania species begin when an infected Phlebotomine bites and injects promastigotes in the skin of the mammal host. Those promastigotes are engulfed by phagocytic cells, as macrophages and dendritic cells. The parasites are kept inside in a parasitophorous vacuole, then they will transform to amastigotes and divide until brock the cell membrane. They are released to infect new cells as amastigotes stage. When an uninfected sand fly bites an infected mammal reservoir, the sandfly ingests the amastigotes, therefore they transform in promastigotes and divide in the midgut of the fly, those promastigotes migrates to the proboscis and are able to produce Leishmaniasis disease. There are no blood stages in the life cycle of L. mexicana (unlike Malaria and Trypanosomiasis).

L. mexicana can induce the cutaneous and diffuse cutaneous clinical manifestations in humans. The cutaneous type develops an ulcer at the bite site, here the amastigotes do not spread and the ulcers become visible either a few days or several months after the initial bite. The diffuse cutaneous type begins when the amastigote diffuses through the skin and metastasize to other tissue. This type does not produce ulcers and there is no treatment.

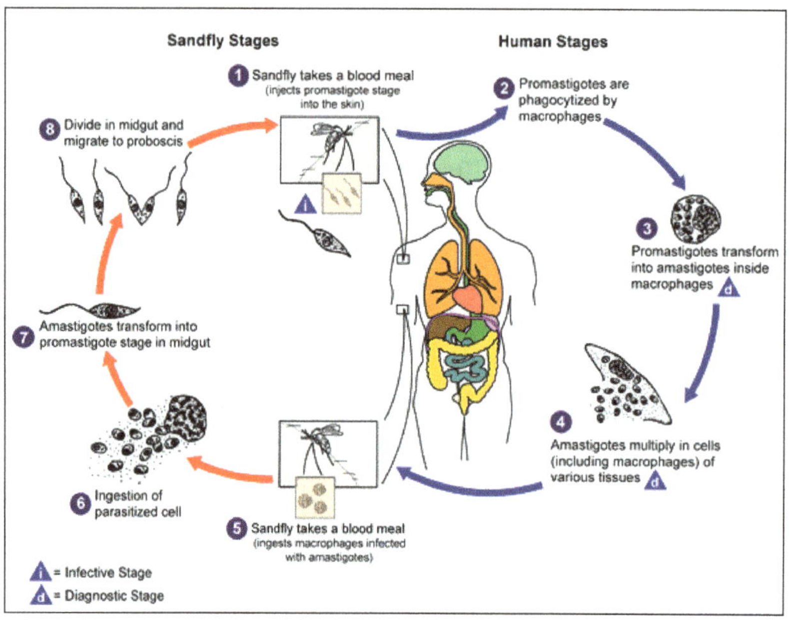

Treatment of Leishmaniasis caused by L. mexicana consists on pentavalent antimonials as Pentostam or Glucantime injected direct into the ulcer or Intramuscular.

Prevention of L. mexicana infection is principally avoiding the bite of the infected sandfly.

Leishmania Major

Leishmania major is a species of parasites found in the genus Leishmania, and is associated with the disease zoonotic cutaneous leishmaniasis (also known as Aleppo boil, Baghdad boil, Bay sore, Biskra button, Chiclero ulcer, Delhi boil, Kandahar sore, Lahore sore, Oriental sore, Pian bois, and Uta). L. major is an intracellular pathogen which infects the macrophages and dendritic cells of the immune system. Though Leishmania species are found on every continent aside from Antarctica, Leishmania major is found only in the Eastern Hemisphere, specifically in Northern Africa, the Middle East, Northwestern China, and Northwestern India.

Life Cycle

As a trypanosomatid, L. major begins its lifecycle in promastigote form in the midgut of the main vector, female sand flies (Phlebotomus spp.). Once in the gut of the sand fly, the parasites change from aflagelated amastigotes into flagellated promastigotes for 1–2 weeks until they are fully developed, a which point they make their way to the proboscis. Upon biting a mammalian host, promastigotes are released into the bloodstream, where they are engulfed by macrophages. Following engulfment, promastigotes differentiate into amastigotes. Amastigotes are oval or round, and have a diameter between 2-3μm. Additionally, they contain a large, eccentrically placed nucleus along with a kinetoplast (which holds extracellular DNA). Being equipped to survive the acidic environment inside the phagosomes of macrophages, the amastigotes reproduce through the process of binary fission. At this point the amastigotes are released throughout the body, and can be ingested by female sand flies, thus completing the cycle. L. major has a sexual cycle, including a meiotic process. Mating only occurs in the sand fly vector.

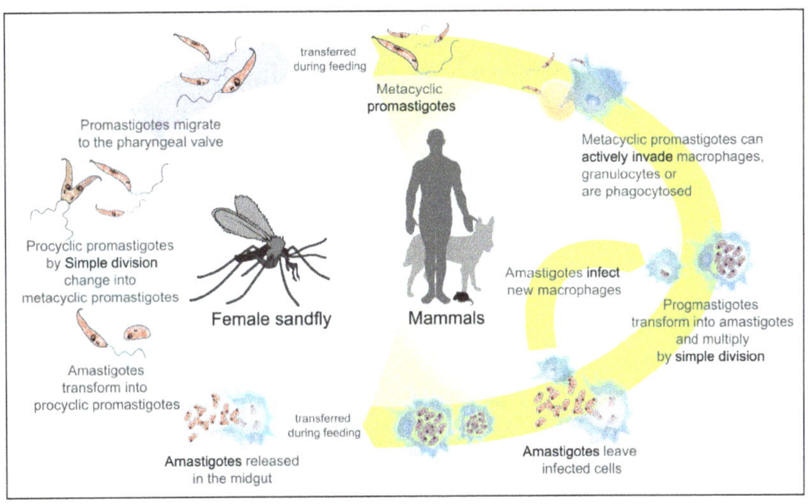

Leishmania major Life Cycle.

Infection

Upon entering the mammalian bloodstream, L. major meets the focal point of infection, the macrophage. As a result of two surface molecules, the protease gp63 and a lipophosphoglycan, promastigotes are able to bind to several macrophage receptors. Promastigote attachment to macrophages is facilitated by a number of receptors, including complement receptors CR1 and CR3, and the receptor for advanced glycosylation end products. Activation of complements occurs far from the cell membrane, and insertion of the membrane attack complex does not occur. This action is what allows the parasite to avoid being lysed, and to persist within the host's macrophages.

Epidemiology

The incidence rate of cutaneous leishmaniasis is estimated to be between 1-1.5 million cases a year. However, transmission does not often occur in utero, during blood transfusions, or through interpersonal contact. Thus, the main form of transmission is through the sand fly vector. Sand flies do not fly long distances, and tend to complete their life cycles in areas with a diameter of less than 1 km. Furthermore, because of the propensity of sand flies to seek out shelter in the burrows of small rodents, where L. major is endemic, small mammals such as gerbils and birds serve as the main reservoirs. Dogs have also been documented as contracting cutaneous leishmaniasis in Egypt and Saudi Arabia. This is rare however, and dogs are not important hosts for L. major.

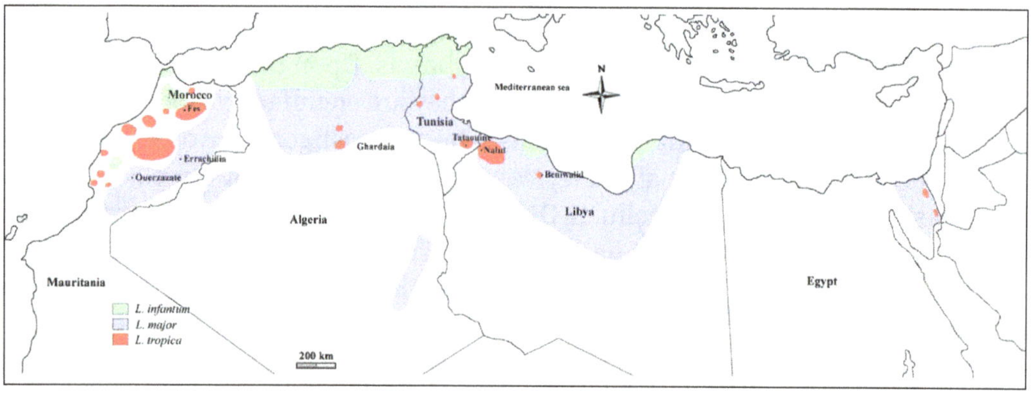

Cutaneous leishmaniasis in North Africa; Leishmania major infected areas are in blue

L. major and its cousin, L. tropica, are recognized as causing the majority of cases of cutaneous leishmaniasis across the Middle East, Northern Africa, and some areas of China and India (as mentioned above). Between 2002 and 2004, over 700 cases of the disease were reported among United States military personnel serving in Iraq.

Clinical Manifestations

Upon becoming infected, patients usually present with lesions at the site of the sand fly bite. The infection is acute, and usually has a duration of about 3–6 months. As more and more phagocytic cells engulf promastigotes, prompting the production of amastigotes, nodules form on the skin. These nodules then ulcerate, although due to the variable characteristics of the lesions, species specific identification of the pathogen is impossible. Generally though, lesions appear moist and have raised outer borders, a granulating base, an overlying layer of white purulent exudate, and have been described as "pizza-like." Biopsies of these lesions usually reveal a number of findings including numerous macrophages containing intracellular amastigotes as well as lymphocytes with observed granuloma formation and few parasites.

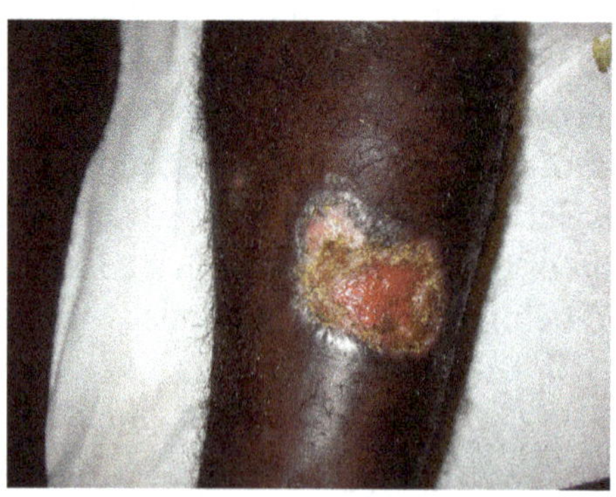

Cutaneous Leishmaniasis Sore

Diagnosis

L. major should be considered in the differential diagnosis of chronic lesions of people who have spent time in areas where it is endemic. However, other pathogens can cause similar lesions and therefore paracoccidiodomycosis, histoplasmosis, sporotrichosis, lobomycosis, lupus vulgaris, Mycobacterium ulcerans, syphilis, cutaneous sarcoidosis, and leprosy should all be considered as well.

The most common ways of diagnosing leishmaniasis are to identify amastigotes in a Wright-Giemsa-stained touch preparation or through isolation of the parasites in cultures.

Treatment

Because the host's immune system tends to resolve infection after 3–6 months, treatment of the lesions generally focuses on limiting tissue damage and necrosis. A number of different treatments have yielded results of varying effectiveness in the treatment of L. major caused cutaneous leishmaniasis.

- Fluconazole given in 200 mg doses over the course of 6 weeks resulted in 90% cure rate versus 60% in those given a placebo.
- Topical application of 15% paromomycin and 12% methylbenzethonium has been used successfully to treat patients in Israel.
- Intralesional injections of 0.5-2.0mL of 100 mg/ML antimony has also been shown to be effective when injected around the sides of lesions. When 10 such injections were given to patients in Egypt, 85% were cured within 3 months.

Prevention

The spread of leishmaniasis can be prevented by interrupting the sand fly life cycle or removing or treating pathogen reservoirs. Avoiding sand fly bites is an effective means of avoiding disease for short term visitors to areas where L. major is endemic. This can be accomplished through the use of DEET containing insect repellent, application of insecticides to clothes and bedding, as well as using mosquito nets to cover beds. Sand flies usually bite between dusk and dawn, so preventative measures should be taken during these times.

Though a vaccine does not yet exist that can prevent cutaneous leishmaniasis, it is speculated that one will be developed in the near future. Patients who have recovered from L. major infections develop high- level immunity to the pathogen. In Russia and Israel, soldiers were "immunized" against L. major through the injection of live promastigotes into the buttocks; however, this form of treatment was discontinued in Israel due to the occasional formation of large or slow-healing lesions.

Entamoeba

Entamoeba is a genus of Amoebozoa found as internal parasites or commensals of animals.

In 1875, Fedor Lösch described the first proven case of amoebic dysentery in St. Petersburg, Russia. He referred to the amoeba he observed microscopically as Amoeba coli; however, it is not clear whether he was using this as a descriptive term or intended it as a formal taxonomic name. The genus Entamoeba was defined by Casagrandi and Barbagallo for the species Entamoeba coli, which is known to be a commensal organism. Lösch's organism was renamed Entamoeba histolytica by Fritz Schaudinn in 1903; he later died, in 1906, from a self-inflicted infection when studying this amoeba. For a time during the first half of the 20th century the entire genus Entamoeba was transferred to Endamoeba, a genus of amoebas infecting invertebrates about which little is known. This move was reversed by the International Commission on Zoological Nomenclature in the late 1950s, and Entamoeba has stayed 'stable' ever since.

Species

Several species are found in humans and animals. Entamoeba histolytica is the pathogen responsible for invasive 'amoebiasis' (which includes amoebic dysentery and amoebic liver abscesses). Others such as Entamoeba coli and Entamoeba dispar are harmless. With the exception of Entamoeba gingivalis, which lives in the mouth, and E. moshkovskii, which is frequently isolated from river and lake sediments, all Entamoeba species are found in the intestines of the animals they infect. Entamoeba invadens is a species that can cause a disease similar to E. histolytica but in reptiles. In contrast to other species, E. invadens forms cysts in vitro in the absence of bacteria and is used as a model system to study this aspect of the life cycle. Many other species of Entamoeba have been described, and it is likely that many others remain to be found.

Structure

Entamoeba cells are small, with a single nucleus and typically a single lobose pseudopod taking the form of a clear anterior bulge. They have a simple life cycle. The trophozoite (feeding-dividing form) is approximately 10-20 µm in diameter and feeds primarily on bacteria. It divides by simple binary fission to form two smaller daughter cells. Almost all species form cysts, the stage involved in transmission (the exception is Entamoeba gingivalis). Depending on the species, these can have one, four or eight nuclei and are variable in size; these characteristics help in species identification.

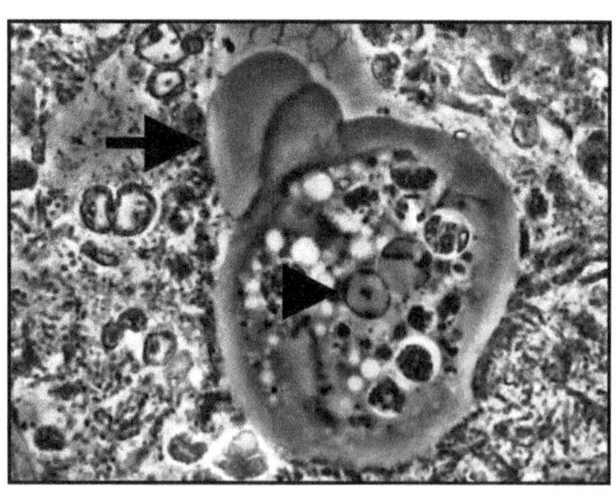

Entamoeba gingivalis.

Classification

Entamoeba belongs to the Archamoebae, which like many other anaerobic eukaryotes lack mitochondria. This group also includes Endolimax and Iodamoeba, which also live in animal intestines and are similar in appearance to Entamoeba, although this may partly be due to convergence. Also in this group are the free-living amoebo-flagellates of the genus Mastigamoeba and related genera. Certain other genera of symbiotic amoebae, such as Endamoeba, might prove to be synonyms of Entamoeba but this is still unclear.

Culture

Fission

Studying Entamoeba invadens, David Biron of the Weizmann Institute of Science and coworkers found that about one third of the cells are unable to separate unaided and recruit a neighboring amoeba (dubbed the "midwife") to complete the fission. He writes:

> When an amoeba divides, the two daughter cells stay attached by a tubular tether which remains intact unless mechanically severed. If called upon, the neighbouring amoeba midwife travels up to 200 μm towards the dividing amoeba, usually advancing in a straight trajectory with an average velocity of about 0.5 μm/s. The midwife then proceeds to rupture the connection, after which all three amoebae move on.

They also reported a similar behavior in Dictyostelium.

Since E. histolytica does not form cysts in the absence of bacteria, E. invadens has become used as a model for encystation studies as it will form cysts under axenic growth conditions, which simplifies analysis. After inducing encystation in E. invadens, DNA replication increases initially and then slows down. On completion of encystation, predominantly tetra-nucleate cysts are formed along with some uni-, bi- and tri-nucleate cysts.

Differentiation and Cell Biology

Uninucleated trophozoites convert into cysts in a process called encystation. The number of nuclei in the cyst varies from 1 to 8 among species and is one of the characteristics used to tell species apart. Of the species already mentioned, Entamoeba coli forms cysts with 8 nuclei while the others form tetra-nucleated cysts. Since E. histolytica does not form cysts in vitro in the absence of bacteria, it is not possible to study the differentiation process in detail in that species. Instead the differentiation process is studied using E. invadens, a reptilian parasite that causes a very similar disease to E. histolytica and which can be induced to encyst in vitro. Until recently there was no genetic transfection vector available for this organism and detailed study at the cellular level was not possible. However, recently a transfection vector was developed and the transfection conditions for E. invadens were optimised which should enhance the research possibilities at the molecular level of the differentiation process.

Meiosis

In sexually reproducing eukaryotes, homologous recombination (HR) ordinarily occurs during meiosis. The meiosis-specific recombinase, Dmc1, is required for efficient meiotic HR, and Dmc1 is expressed in E. histolytica. The purified Dmc1 from E. histolytica forms presynaptic filaments and catalyzes ATP-dependent homologous DNA pairing and DNA strand exchange over at least several thousand base pairs. The DNA pairing and strand exchange reactions are enhanced by the eukaryotic meiosis-specific recombination accessory factor (heterodimer) Hop2-Mnd1. These processes are central to meiotic recombination, suggesting that E. histolytica undergoes meiosis.

Studies of E. invadens found that, during the conversion from the tetraploid uninucleate trophozoite to the tetranucleate cyst, homologous recombination is enhanced. Expression of genes with functions related to the major steps of meiotic recombination also increased during encystations. These findings in E. invadens, combined with evidence from studies of E. histolytica indicate the presence of meiosis in the Entamoeba.

Entamoeba Histolytica

Entameoba histolytica was discovered by a Russian zoologist, Friedrick Losch in 1875. This is an endoparasite common in man, apes and monkeys, and may also be found in pigs, cats and rats. E. histolytica is a microscopic endoparasite and is the causative agent of amoebic dysentery in man. It is seen inhabiting the upper part of large intestine (colon). There are two races or strains of this parasite, a smaller, common, nonpathogenic form and a much larger pathogenic form.

About 600 million people around world are infected by this parasite. Infection of this parasite is high in India, China, Philippines, Thailand, Mexico and parts of South America. A higher prevalence occurs in males, but usually the more serious and fatal complications affect women and children.

Morphology

The most active, motile and feeding form is called trophozoite, and has average about 25 um in diameter. The cytoplasm consists of a clear ectoplasm and a finely granular endoplasm. The endoplasm contain food vacuoles filled with RBC in different stages of digestion. The nucleus is single and vesicular, measures about 3 to 5 um in diameter.

Within it lies the endosome (nucleolus), which may be a single granule or a closely packed cluster of minute granules. A ring appears to surround the endosome. Spoke-like lines radiate from the endosome to the nuclear membrane.

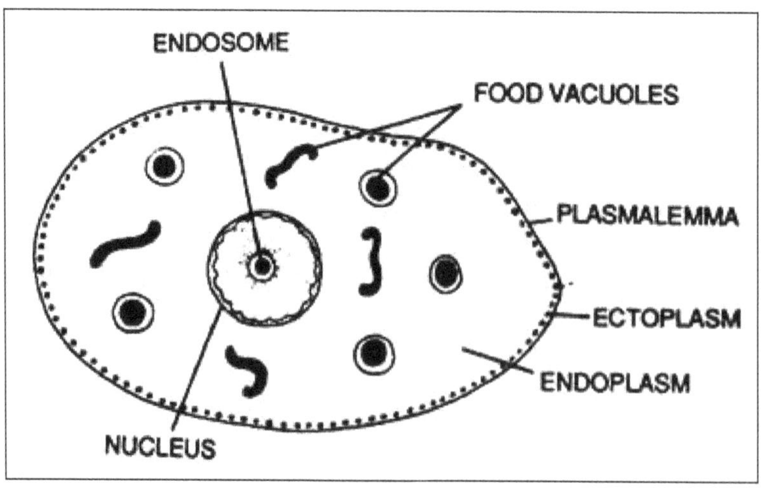

Entamoeba Coli.

Trophozoite produces finger-like pseudopodium and movement is irregular. As there used to be only one pseudopodium, it is typically monopodial. The type and rapidity of movement varies depending upon the consistency of surrounding medium, age of the parasite, temperature etc.

The nutrition in trophozoite is holozoic. It feeds by phagocytosis. Food usually consists of bacteria or other organic material found in the intestine. RBCs are found only in the food vacuole of pathogenic forms.

Pathogenesis

The common, smaller, non-pathogeic minuta form lives in the lumen of the large intestine of man. It is harmless and feeds on bacteria. After a period of feeding and reproduction, vacuoles disappear. The Entamoeba becomes rounded. Soon a cyst wall begins to form, these uninucleated stages are precysts.

If the resistance of gut is lowered in infected people, nonpathogenic form may change to pathogenic magna form. It can also be produced by an alternation of host's diet, increase of host cholesterol, change in the level of host sexual hormones and enzymatic action of the parasite. Magna form is invasive, and it penetrate the intestinal wall by secreting the histolytic enzymes.

They invade the submucosa by dissolving the mucosa of intestinal wall. They multiply by binary fission and produce flask-shaped ulcers containing cellular debris, RBCs, lymphocytes and bacteria. These ulcers rupture and discharge blood and mucus into the intestine that pass to outside with the stool. This results in amoebic dysentery or amoebiasis.

Other symptoms of amoebiasis are vomiting, mild fever, diarrhoea, tenderness over the sigmoidal region of the colon and hepatitis. In the stools of the patient, swarms of Entamoeba and RBCs are found. Stool is usually acidic. The patient is discomforted due to intense abdominal pain with the passing of blood and mucus with stools after every few minutes.

E. histolytica may not be confined to the intestine. Once a lesion is made in the intestinal wall, the amoebae can travel through blood or lymph circulation of host and can invade other tissues like liver, spleen, lungs, skin, gonads and rarely brain.

In all these soft tissues, Entamoeba multiply and cause formation of abscesses. The most common extraintestinal locus of infection in the liver, particularly the right lobe. Hence results in hepatitis. Formation of abscesses in skin result in cutaneous amoebiasis. Abscesses in brain usually result in the death of the host.

Treatment

Many drugs are available for the treatment of amoebic dysentery. Alkaloid Emetine is effective in giving temporary relief. Dehydroemetine, a synthetic derivative of Emetine is equally effective. Chloroquine is used to treat amoebic hepatitis. Other medicine are iodine compounds such as. Chinioform, Diodoquin, Iroform, Humat in, Mexaform and Vioform.

Antibiotics useful in treating amoebiasis are Aureomycin, Fumagillin, Erythromycin and Terramycin. Recently both acute intestinal amoebiasis and amoebic hepatic abscess are treated with Metronidazole (1-(2- hydroxyethyl) -2 methyl 5 nitroimidazole.). Selection of drugs depends on the location of infection and condition of the patient.

Control

1. Hands should be properly washed before handling food materials and eating the food.

2. Finger nails should be cut regularly and kept clean.
3. Uncooked vegetables like salad should be eaten only after through washing.
4. Avoid irrigation of vegetable by contaminated water.
5. Protection of food from contamination from housefly, cockroach etc.
6. Vegetables should be washed before cooking.
7. Avoid defaecation on vegetable fields and open grounds.
8. Quick and proper disposal of night soil.
9. Proper sanitation of surroundings.
10. Chlorination of drinking water supply.
11. Proper disposal of sewage.
12. Chemical treatment of human faeces to be used as fertilizer.
13. Covering of eatables by traders.
14. Killing of cysts with alcohol, phenol, formalin, mercuric chloride hydrochloric acid, caustic soda and potassium permanganate.

Entamoeba Coli

Entamoeba coli are a protozoan endocommensel, inhabiting the lumen of the large intestine of man. There is no reliable evidence that it produces disease in human beings but few workers have reported ingestion of red blood cells by the organism. E. coli was discovered in India by Lewis in 1870 however its detail description was given by Grassi (1879).

Geographical Distribution

It is cosmopolitan in distribution and has been stated to occur in about 50% of human population.

Life Cycle

Entamoeba coli are a monogenetic organism. Three distinct morphological forms exist airing the life cycle-Trophozoite, Pre-cystic stage and Cystic stage.

Irophozoite of E. coli is about 20 to 30 in diameter with a range from 10 to 50. Trophozoite is unicellular. The cytoplasm is differentiated into outer narrow ectoplasm which is not so prominent and inner granular, vacuolated endoplasm containing bacteria and debris inside food vacuoles.

A single nucleus lies inside the endoplasm. The nucleus is a ring like structure with thick nuclear membrane lined with irregularly distributed masses of chromatin and a large, irregular, eccentric karyosome.

Fine linin threads extend between nuclear membrane and karyosome. Trophozoite bears one too many pseudopodia which are short, blunt and granular Movement is sluggish and usually not directional. The parasite feeds upon bacteria, vegetable cells and other faecal debris present in the large intestine. Dobell (1938) reported that it may ingest R.B.C., occasionally. The trophozoite rcproduces by binary fission.

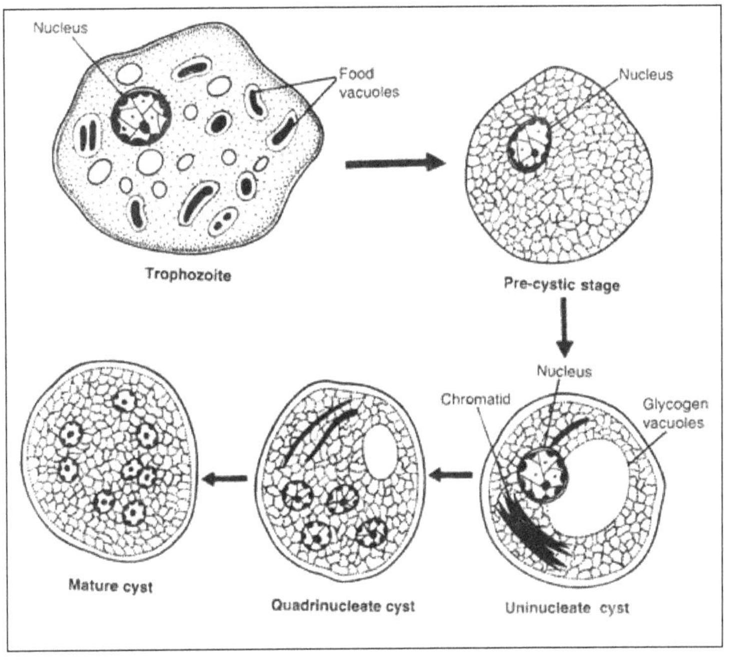

Trophozoite changes into spherical uninucleate precystic stage. The precystic stage size ranges from 15 to 45 μ in diameter. It is similar to trophozoite stage, except that it is non feeding stage and hence food inclusions are not found in the endoplasm. Precystic stage changes into cystic stage.

The cysts are spherical or avoid with size ranging from 10 to 33 μ in diameter. The cyst wall is thick. Immature cyst may have one-two or four nuclei with eccentric kaiyosome .Occasionally, the cyst may bear 16 or even 32 nuclei. Glycogen vacuoles and chromatid bodies are seen in the endoplasm up to binucleate stage after that they are consumed.

Matured cyst is the infective stage. Cyst formed in the large intestine is discharged out o the host's body through faeces. The cysts survive for 3-4 months outside the body of the host and are relatively more resistant to desiccation as compared to those of E. histolytica. The survive rate of the cyst is about 46 per cent.

Mode of Infection

Infection to the new host occurs by consuming contaminated food and drinks. The infective stage cysts are carried from faces to the food items through insects and rodents. In the small intestine of the new host excystation occurs during which a single multinucleate amoeba comes out through the cyst wall. Multinucleate amoeba divide into as many immature amoebas as there are nuclei in the cyst.

The young amoeba moves down to reach the caecum where they multiply in number and become trophozoites.

Pathology

E. coli lives inside the lumen of the large intestine in man. They never enter into the mucosa or sub-mucosa layers or other tissues of the intestine. There is no reliable evidence that it ever produces intestinal lession, although it has been reported that E. coli occasionally ingest red blood cells.

In this way it is believed to exist as non-pathogemc endo-commensa. However, Dey (1974) observed that a large population of E. coli inside the gut lumen may cause dyspepsia, hyperacidity, gastritis and indigestion.

Treatment

Since, Entamoeba coli is harmless, its treatment is not required from the stand point of health, however arsenical compounds such as carbarsone have been reported to be quite effective to kill the trophozoites.

Prophylaxis

The prophylaxis of E. coli is similar to that of E. histolytica.

Entamoeba Hartmanni

Entamoeba hartmanni is a non-pathogenic amoeba with worldwide distribution. Its life cycle is similar to that of E. histolytica but it does not have an invasive stage and does not ingest red blood cells.

Morphology of Cysts

Cysts of E. hartmanni 7-9µ in diameter and contain 1 - 4 nuclei. Chromatoid bodies are usually present in young cysts as elongated bars with bluntly rounded ends. Glycogen is usually diffuse, but in young cysts it is often present as a concentrated mass, staining reddish brown with iodine.

Morphology of Trophozoites

Morphology of the trophozoites is similar to those of E. histolytica/dispar but they do not contain ingested red blood cells and the motility is less rapid.

Laboratory Diagnosis

Laboratory diagnosis is made by finding the characteristic cysts in an iodine stained, formol-ether concentration method or by detecting the characteristic trophozoites in a wet preparation or a permanent stained preparation.

Entamoeba Polecki

Entamoeba polecki is an intestinal parasite of the genus Entamoeba. E. polecki is found primarily in pigs and monkeys and is largely considered non-pathogenic in humans, although there have been some reports regarding symptomatic infections of humans. Prevalence is concentrated in New Guinea, with distribution also recorded in areas of southeast Asia, France, and the United States.

Morphology

Mature trophozoites of E. polecki are generally 10-20 μm in diameter. Trophozoites are irregularly shaped and possess pseudopodia for motility. They have a single nucleus with a small central karyosome and finely dispersed peripheral chromatin, similar to that of Entamoeba histolytica. Cytoplasmic contents are similar to other Entamoeba sp. and are usually granular and vacuolated. Cysts of E. polecki are morphologically unique, containing only one nucleus, varyingly sized chromatoid bars, and a large inclusion mass.

Transmission and Life Cycle

There are two stages in the life cycle of E. polecki. The first is as a trophozoite, a vegetative stage that cannot survive in the environment. The second is a cyst, where transmission of parasite is possible and provides protection to harsh external environments. Cysts are infective when ingested by another organism. The cystic form of this protozoan has a diameter as small as 9.5 μm and as large as 17.5 μm. Morphologically, E. polecki is extremely similar to Entamoeba histolytica and Entamoeba hartmanni.

Transmission follows a fecal-oral route. Infected feces with mature cysts are ingested where the cyst matures to the trophozoite in the gastrointestinal tract of the host. It is considered to be a zoonotic parasite, as close contact with infected swine have been reported to be the cause of E. polecki infections in humans. Transmission to humans from consumption of pork is unlikely. Recent studies suggest that different subspecies infect non-human primates and pigs, and close inhabitation between the two do not coincide with transmission.

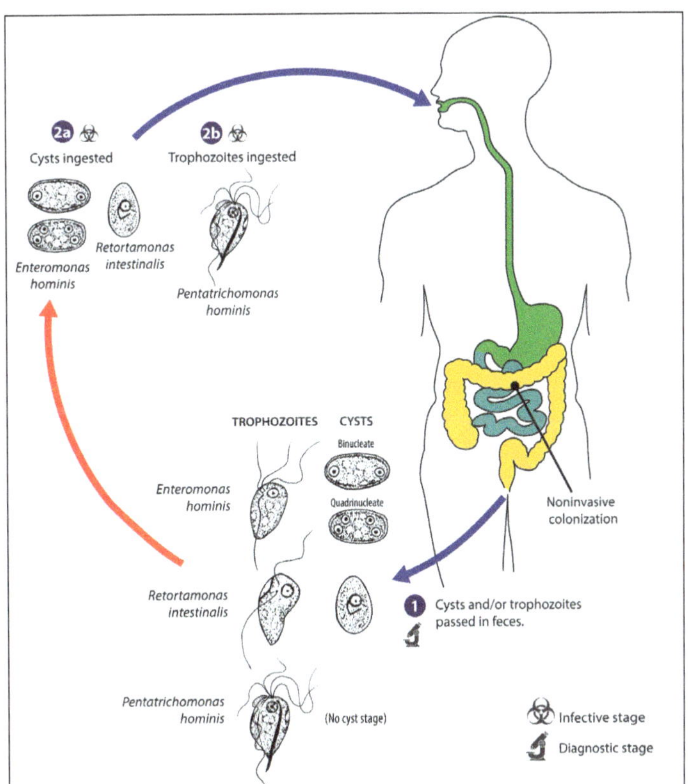

Life cycle of Entamoeba sp.

Pathology

E. polecki is considered to be non-pathogenic in humans. Nonspecific symptoms from infection have been reported, including nausea, vomiting, diarrhea, bloody stools, and fever. Prevalence of infection amongst swine may be up to 25% across the world, but tend to be asymptomatic.

Diagnosis and Treatment

Examination of stool samples for uninucleated cysts and trophozoites have been used for diagnosis. This method is not always reliable due to morphological similarities between E. polecki and other Entamoeba species. More recent diagnostic methods utilizing DNA amplification and comparison have been used to better differentiate amongst pathogenic and species such as E. histolytica and non-pathogenic species. A definitive diagnosis can be made by using electroimmunotransfer blots. Serological testing is not accurate between species of Entamoeba.

Treatment of infection is similar to that of other Entamoeba infections. Anti-parasitic medications such as metronidazole and ornidazole are generally used to treat human infections. Combination therapies such as metronidazole and diloxanide furoate have been effective as well.

Entamoeba Gingivalis

Entomoeba gingivalis is a protozoan endoparasite, residing in the tartar and puspockets of teeth of pyorrhoea infected human beings. It is the first parasitic amoeba known to human beings. E. gingivalis was first of all observed by Gros in 1849 in the tartar of teeth but its detail description was given by Von Prowazak in 1904. Smith and Barrett (1915) described it as causative agent of pyorrhoea alveolaris.

Geographical Distribution

E. gingivalis is cosmopolitan in distribution. It is estimated that in India more than 70 per cent population is infected by this parasite. With advancing age, the percentage of individuals suffering from the infection of E. gingivalis increases.

Life Cycle

E. gingivalis is a monogenetic parasite. Human beings are their only host, however occasionally the parasite has also been reported from the mouth of dogs, cats, horses and captive monkeys. Only trophozoite stage exists during the life cycle. The size of the trophozoite ranges from 5 to 30 diameters however, the usual size is from 10 to 20 μ.

The single celled trophozoite is differentiated into outer clear ectoplasm and inner granular endoplasm. During non motile condition, ectoplasm is hardly visible but during motile stage it appears as a thick layer comprising about one half of the volume of the actively motile animal.

The endoplasm is granular, vacuolated and usually crowded with floating food particles. Food vacuoles contain dark-staining rounded bodies largely derived from the nuclei of degenerated epithelial cells, lymphocytes and occasionally leukocytes. The organism also, ingests bacteria but do not take red blood cells. E. gingivalis is basically a scavenger of disintegrated cells. Bacteria are a minor source of nourishment.

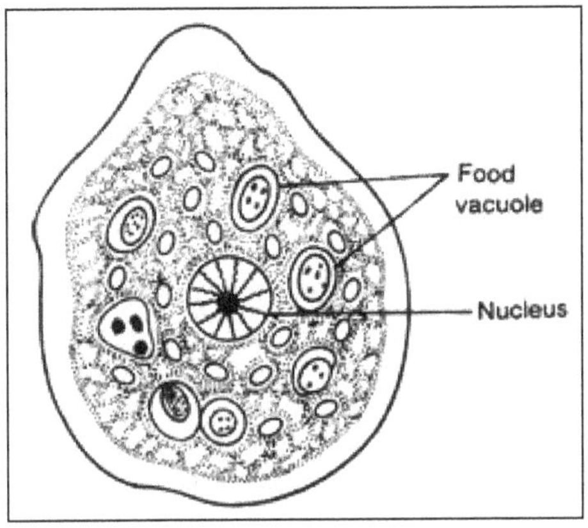

The endoplasm contains a single, small nucleus. The nucleus is spherical and m unstained condition is usually inconspicuous. The moderately thick nuclear membrane contains irregularly distributed small masses of chromatin. A central or eccentric karyosome is present inside the nucleus from which delicate radiating fibrils extends up to peripheral ring.

The parasite moves by forming ectoplasmic pseudopodia. During resting stage several pseudopodia may be seen extruding in various directions but during directional progress one more large pseudopodia ma, be extended. E. gingivalis reproduoes by binary fission. Cysticsrnge have not been seen during the life cycle. Trophozoite donot survive outside the host body.

Mode of Infection

Transmission of E. gingivalis from one person to another occurs by mouth to mouth kissing. It may be transmitted due to coughing by infected cook at the time of dishing.

Pathology

The actual pathogenicity of these protozoa is not well as certain as there is no convincing evidence that the organism causes oral disease. The parasite lies in pyorrhoea alveolaris and was previously supposed to cause pyorrhoea but nowadays, few workers have that suggested though this protozoon actually thrives as a commensal in the gingival tissues of man, it purely prepares the teeth for pyorrhoea and its purely commensal relationship with man is in doubt.

Treatment

Treatment of abnormal oral conditions or disease is the best way to eliminate the parasite. No specific drug or medicine is prescribed to kill the organism.

Prophylaxis

Since, the organism is not directly causing harm to the host; there is no specific prophylaxtic measure to prevent its occurrence. Proper hygiene may reduce the incidence of E. gingivalis.

References

- Protozoan, science: britannica.com, Retrieved 3 March, 2019

- Roberts, Larry S.; Janovy, Jr., John (2009). Foundations of Parasitology (8th ed.). Mcgraw-Hill. Pp. 176–7. ISBN 978-0-07-302827-9
- Leishmania, science: britannica.com, Retrieved 4 April, 2019
- Van Griensven, Johan; Diro, Ermias (June 2012). "Visceral Leishmaniasis". Infectious Disease Clinics of North America. 26 (2): 309–322. Doi:10.1016/j.idc.2012.03.005. PMID 22632641
- Species-of-entamoeba-with-diagram-microbiology, micro-biology: biologydiscussion.com, Retrieved 5 May, 2019
- Parasite-entamoeba-coli-life-cycle-mode-of-infection-and-treatment, zoology: yourarticlelibrary.com, Retrieved 6 June, 2019
- Macpherson CN, Meslin FX, Wandeler AI, eds. (2000). Dogs, Zoonoses and Public Health. New York, NY: CABI Publishing. p. 139. ISBN 0-85199-436-9.
- Entamoeba-hartmanni, Amoeba, Protozoa, faecalparasitology, parasitology: ukneqasmicro.org.uk, Retrieved 7 July, 2019
- Solaymani-Mohammadi, S.; Petri, W.A. (2006). "Zoonotic implications of the swine-transmitted protozoal infections". Veterinary Parasitology. 140 (3–4): 189–203. Doi:10.1016/j.vetpar.2006.05.012. PMID 16828229
- Parasite-entamoeba-gingivalis-life-cycle-mode-of-infection-and-treatment, zoology: yourarticlelibrary.com, Retrieved 8 August, 2019

4

Parasitic Worms

Parasitic worms are macroparasites such as zoonotic hookworms, threadworms, filarial worms, giardia lambia, cestoda, human whipworm and ascaris lumbricoides. They can be either endoparasitic or ectoparasitic. This chapter has been carefully written to provide an easy understanding of these types of parasitic worms.

Zoonotic Hookworms

Hookworms are parasitic intestinal nematodes, several of which are zoonotic. In their normal hosts, hookworms may enter the body either by ingestion or through the skin. Larvae that penetrate the skin travel through various organs, including the respiratory tract, before entering the intestines and developing into mature hookworms. Hookworms can cause anemia, abdominal pain and diarrhea when they reside in the intestines, or respiratory, dermatologic and other signs during their migration through the body. Young individuals tend to be affected more severely. In cattle, infections may lead to severe disease and pronounced weight loss, with as few as 50 adult worms causing significant anemia in calves. Hookworm disease in cats and dogs can result in anemia, and infections of neonatal pups may prove fatal, even with as few as 50-100 worms present.

Animal hookworm larvae can penetrate the human epidermis, but most species cannot readily enter the dermis, and remain trapped in the skin. These larvae migrate extensively within the skin for a time, resulting in a highly pruritic but self-limited disease called cutaneous larva migrans. One species carried by dogs and cats is increasingly recognized as an intestinal parasite of humans: Ancylostoma ceylanicum has been found in 6-23% of patent human hookworm infections in some parts of Asia. A. caninum also migrates occasionally to the intestines, but usually as a single worm. While one hookworm is unlikely to cause significant blood loss, its presence may result in a painful intestinal disorder called eosinophilic enteritis.

Etiology

Hookworms are nematodes in the superfamily Ancylostomatoidea. In their normal hosts, they are parasites of the intestinal tract. Humans are usually infected by Ancylostoma duodenale and Necator americanus, which are maintained in human populations. Some zoonotic species may also reach the intestines. A. ceylanicum can sometimes be found in large enough numbers to cause typical enteric signs, but A. caninum seems to occur only as a single worm. Rare human intestinal infections with A. malayanum, A. japonica, Necator suillis and N. argentinus have also been reported, but the identification of these organisms is uncertain.

Hookworm larvae that normally mature in the intestinal tracts of animals can cause cutaneous larva migrans in people. Zoonotic hookworms known to cause this condition include A. braziliense, A. caninum, A. ceylanicum, A. tubaeforme, Uncinaria stenocephala and Bunostomum phlebotomum. Other species of hookworms found in animals, including wildlife and captive exotics, might also be able to cause cutaneous larva migrans.

Species Affected

Ancylostoma braziliense is a hookworm of dogs, cats and other carnivores. A. caninum is found in dogs, and has been reported in cats in some parts of Asia and Australia. Uncinaria stenocephala also infects dogs and occasionally cats. A. ceylanicum occurs in wild and domesticated canids and felids, and A. tubaeforme infects cats and other felids. Bunostomum phlebotomum is a hookworm of cattle. Rodents can be paratenic hosts for hookworms including A. braziliense, A. tubaeforme, and U. stenocephala, and possibly other hookworm species.

Ancylostoma duodenale and Necator americanus, species usually found only in humans, have been reported on rare occasions in other mammals.

Zoonotic Potential

A. braziliense is responsible for most cases of cutaneous larva migrans in humans. A. caninum, A. ceylanicum, Uncinaria stenocephala and Bunostomum phlebotomum are involved less frequently, while rare cases have been caused by A. tubaeforme.

A. ceylanicum is the only zoonotic hookworm known to produce patent intestinal infections in humans. A. caninum can cause eosinophilic enteritis, but does not seem to become patent.

Geographic Distribution

A. caninum is the most widespread of all hookworms and can be found in many parts of the world. A. tubaeforme is also widely distributed. A. braziliense is limited to tropical and subtropical regions including Central and South America, the Caribbean and parts of the U.S. A ceylanicum has been reported in parts of Asia, Africa, Australia, and the Middle East, and in one publication from Brazil. B. phlebotomum is a parasite of temperate regions, while U. stenocephala occurs in colder climates including Canada, the northern U.S. and Europe.

Transmission and Life Cycle

Adult hookworms live in the intestines. Hookworm eggs shed in the feces are not immediately infective, and hatch in the environment, often within one to a few days. The larvae feed on soil bacteria and molt twice before they become infective third stage larvae. The larvae develop best in warm, moist, sandy soil that is sheltered from direct sunlight. Under optimal conditions, they reach the infective stage in approximately 4 to 7 days. Third-stage larvae that are unable to enter a mammalian host die in approximately 1 to 2 months when their metabolic reserves are exhausted.

Larvae may enter the body either by penetrating the skin, or by ingestion. Penetration of the skin by third stage larvae usually requires at least 5 to 10 minutes contact with contaminated soil. In their natural hosts, these larvae enter the dermis, where they are transported through the lymphatic vessels and veins to the lungs. In the lungs, they penetrate the alveoli and migrate up the respiratory tree to the trachea. They are swallowed and mature into adults in the intestines. Ingested hookworm larvae do not follow this route, but they develop for a period of time in the gastrointestinal wall before re-appearing in the lumen and maturing to adult worms. A. caninum, A. braziliense, A. ceylanicum and B. phlebotomum can either penetrate the skin or be ingested. U. stenocephala is usually acquired by ingestion.

In dogs more than three months old, A. caninum larvae may fail to complete the migration through the lungs and are arrested in the tissues, where they survive as dormant (hypobiotic) larvae. These larvae can move to the uterus or mammary gland during pregnancy, and are transmitted to the pups. This route of transmission does not seem to exist for A. braziliense or U. stenocephala in dogs, or for any hookworms in cats or cattle.

In the intestines, adult hookworms are attached to the mucosa, but change their location every few hours, leaving tiny, bleeding mucosal ulcerations behind. Some species such as A. caninum release a strong anticoagulant that can cause profuse bleeding. Adult hookworms can live for months to a year or more. Reported adult lifespans are approximately 4 to 8 months for A. braziliense, 18 months to 2 years for A. tubaeforme, and about 4 months to a year for U. stenocephala in dogs. The prepatent period varies with the species of parasite, host species and route of exposure. Canine and feline hookworms usually become patent after 13-27 days; however, A. caninum infections transmitted in colostrum or in utero can produce eggs during the second week of life. The prepatent period for B. phlebotomum in cattle is approximately two months. Eggs may be shed intermittently.

Infections in Paratenic Hosts

Paratenic (transport) hosts can be infected orally or through the skin. Larvae do not develop further in paratenic hosts, but become dormant in various tissues. In mice, A. braziliense and A. tubaeforme larvae are mainly found in the head, particularly in the nasopharyngeal epithelium and salivary glands. A. caninum and U. stenocephala occur mainly in the muscles. If a definitive host ingests these larvae, they are released and complete their development to adults.

Infections in Humans

In humans, most zoonotic hookworm larvae cannot penetrate into the dermis. They remain confined to the epidermis, where they migrate for a period of time but eventually die. These organisms cannot be transmitted to others.

A. caninum can occasionally be found in human intestines, while A. ceylanicum seems to occur more frequently. The route of infection with A. caninum is still unknown; either percutaneous or oral transmission may be possible. Infections with this organism do not seem to become patent; only single A. caninum worms have ever been found in humans. A. ceylanicum, however, can mature, mate and produce eggs in people. Patients with intestinal hookworm disease (A. ceylanicum) are not directly contagious to others, as the eggs must develop for a period of time in the soil before they develop into infective third stage larvae. However, the eggs they shed can contaminate the soil. Epidemiological and genetic data suggest that A. ceylanicum may cycle between humans, dogs and cats in some hookworm-endemic communities in Southeast Asia. It is possible that this occurs in other locations as well.

Disinfection

Aqueous iodine at 50-60 parts per million at 15-30 °C (59-86 °F) has been reported to kill hookworm larvae in 15 minutes or less. Other agents which were shown to kill larvae of Necator americanus in a laboratory setting include very hot water (above 80 °C, 176 °F), ethanol (70% for 10 minutes contact time), and Dettol (0.5% for 15 minutes contact time). In the same study, sodium hypochlorite (bleach, 1%) and glutaraldehyde (2%) had no killing effect on N. americanus larvae, and the efficacy of these agents on the larvae of other hookworm species is questionable. Sodium borate (1kg/2 m2) can be used to disinfect the soil. Hookworm larvae are also susceptible to freezing, drying, direct sunlight and temperatures above 45 °C (113 °F).

Infections in Animals

Incubation Period

The incubation period varies with the number of parasites. Puppies can become symptomatic in the first week of life, before the infection becomes patent.

Clinical Signs

Dogs and Cats

The clinical signs caused by adult hookworms vary with the parasite burden and the age of the animal. They are frequently related to enteritis and/or intestinal blood loss, and are generally more severe in young animals.

In dogs, A. caninum can cause anemia, dark reddishbrown to black hemorrhagic diarrhea, anorexia and dehydration, with associated weakness. Death may occur due to blood loss. The worms can also cause protein and fluid loss and malabsorption, resulting in wasting and decreased growth. Older animals can carry a few worms without clinical signs. Similarly, A. tubaeforme can cause intestinal blood loss, anemia and weight loss in kittens, and large numbers of worms can be fatal. In contrast, U. stenocephala and A. braziliense are not heavy blood-feeders and do not cause anemia or bloody diarrhea. However, they can result in enteric disease, including diarrhea and proteinlosing enteropathy.

Larval hookworms may also cause clinical signs during their migration. Dermatitis may be seen where they penetrate the skin. The cutaneous lesions, which can include erythema, pruritus and papules, are usually limited to the feet and often to the interdigital spaces. In some cases, these signs may be severe and result in self-inflicted trauma. Most often, the skin lesions disappear approximately five days after they appear. Large numbers of larvae in puppies can cause pneumonia during their migration through the lungs. Rarely, aberrant larvae in other locations (e.g., the spinal cord) may also become symptomatic.

Cattle

Larval penetration of the lower limbs can cause uneasiness and stamping, and there may be local skin lesions, edema and scabs. The adult worms can cause anemia, rapid weight loss, and alternating diarrhea and constipation. Hypoproteinemia may be seen, but bottle jaw is usually mild. Deaths can occur, especially in calves.

Post Mortem Lesions

Hookworms are small, grayish-white to reddish-white, cylindrical nematodes (approximately 5-20 mm long), found in the intestines. The intestinal mucosa may be congested and swollen, with many tiny hemorrhagic points or ulcers. With many hookworm species, the intestinal contents are bloodstained. In animals with anemia, the liver and other organs may appear pale.

Pneumonia and lung consolidation can be seen with large numbers of larvae in puppies. Skin lesions may be found on the feet, particularly between the toes where the larvae penetrated. Larvae in aberrant sites may be associated with necrotic and hemorrhagic tracts in the tissues, as well as other signs of tissue damage.

Diagnostic Tests

Hookworm infections are diagnosed by fecal flotation and detection of the eggs. Typical Ancylostoma eggs are 55 – 76 µm in length and approximately 34 – 50 µm in width, and have a smooth, thin outer shell. They are unembryonated when they are first shed, but develop quickly; at the time of diagnosis, they may contain several cells or a ball of cells. Uncinaria spp. eggs are very similar but slightly larger (70 – 90 µm × 40 – 50 µm); they cannot be easily distinguished from Ancylostoma spp. except in mixed infections. Firststage hookworm larvae may appear in preparations from old or stored feces, especially in warm and humid conditions. Eggs are not shed constantly, and repeated sampling may be necessary to detect infections. Coproantigen ELISA tests are in development.

Hookworm larvae can be identified by fecal culture, if identification to the species level is important. Adult worms can be differentiated by their morphology, using published keys. Polymerase chain reaction (PCR) assays have been used in research.

Treatment

Hookworms can be treated with a wide variety of anthelmintics, however, resistance has been detected in the case of some commonly used drugs such as pyrantel in dogs. Supportive care such as supplemental iron, blood transfusions or a high protein diet may also be necessary in some cases.

Prevention

Anthelmintics can be used in ruminants to decrease parasite burdens and pasture contamination. Pasture rotation and other management techniques can also be important components in preventing disease.

Some heartworm preventives may also aid in the prevention of hookworm disease in dogs and cats. Concrete runways, washed at least twice a week in warm weather, should be used for dogs housed in kennels. Clay or sandy runways, as well as soil and lawns, can be decontaminated with sodium borate. To prevent A. caninum infections in puppies, bitches should be free of hookworms, and they should be kept out of contaminated areas during their pregnancy. The dam and puppies should also be housed separately from other animals.

Infections in cats can be decreased by keeping cats indoors and preventing them from eating rodents. Keeping the litter box clean may decrease reinfection.

Morbidity and Mortality

Hookworms are common parasites of cats and dogs. Although their prevalence can vary, species such as A. caninum or A. tubaeforme may infect most of the dogs and cats in some tropical regions. A. ceylanicum is also common in some communities in Asia. In one study, A. ceylanicum was found in 46% of hookworm-infected dogs and cats in a rural community in Malaysia, where nearly 25% of hookworm-positive humans were found to be infected with this same species.

A. caninum and A. tubaeforme infections are generally more serious than U. stenocephala and A. braziliense. Some A. caninum and A. tubaeforme infections may be fatal, particularly in young animals, due to blood loss. B. phlebotomum can cause deaths in calves.

Infections in Humans

Incubation Period

The incubation period for cutaneous larva migrans is short but vaguely established; according to some estimates, it is approximately 1 to 2 weeks. The incubation period for intestinal hookworm disease varies with the number of parasites and can be a few weeks to many months.

Clinical Signs

Cutaneous Larva Migrans

Cutaneous larva migrans is the most common syndrome caused by zoonotic hookworms in humans. Most lesions occur on the legs, buttocks and hands, but they can be found on any part of the body that was exposed to the soil. Initially, there may be a tingling or prickling sensation where the larvae penetrated the skin, followed by a papule at the same location.

Migration of the slow-moving larvae in the skin results in an allergic reaction where they tunnel. The lesions may include papules as well as nonspecific dermatitis, vesicles, or narrow, serpiginous (snakelike), slightly elevated, erythematous lines. The lesions are intensely pruritic, especially at night, and usually advance several millimeters to a few centimeters a day. Pain is occasionally reported, usually in association with vesicles. Secondary bacterial infections can occur due to scratching. Most cases resolve spontaneously in a few days to several weeks, but some untreated lesions have been reported to last for more than a year.

Other lesions are occasionally reported, when larvae penetrate beyond the epidermis. A. caninum larvae may migrate to the muscles, resulting in myositis with persistent swelling and tenderness. These larvae can also cause systemic signs and folliculitis. Ancylostoma spp. Larvae have been documented in the eye.

Classic (Intestinal) Hookworm Disease

Although intestinal hookworm disease is usually caused by human hookworms, the zoonotic species A. ceylanicum can also cause this syndrome. With hookworms adapted to humans, the first symptom is usually pruritus at the site of larval penetration. There may also be erythema with small papules or vesicles, which usually persists for 1 to 2 weeks. Migration of the larvae through the lungs may cause coughing and wheezing; however, lung signs are uncommon and are usually mild except with very heavy worm burdens. The adult worms can cause acute intestinal signs such as abdominal pain, nausea, anorexia, vomiting and hemorrhagic diarrhea or melena. Chronic hookworm disease is characterized by blood loss and iron-deficiency anemia, and is associated with fatigue, pallor, tachycardia and dyspnea on exertion. Hypoproteinemia may cause edema, and there can be signs of malabsorption and malnutrition. In children, there may also be adverse effects on physical and intellectual growth. The severity of the disease varies with the worm burden and the amount of blood loss. Heavy infections can be fatal, particularly in infants.

A. ceylanicum infections tend to be milder than those caused by human hookworms. Anemia is usually the most prominent symptom, but other clinical signs, similar to those caused by the human hookworms, are also reported.

Eosinophilic Enteritis

Eosinophilic enteritis is caused by the zoonotic hookworm A. caninum. It is characterized by increasingly severe episodes of abdominal pain associated with peripheral eosinophilia, but no blood loss. Severe cases can mimic appendicitis or intestinal perforation. Some A. caninum infections may be asymptomatic.

Diagnostic Tests

A presumptive diagnosis of cutaneous larva migrans is usually made based on the clinical signs. It may be possible to confirm the diagnosis by a biopsy of the affected skin; however, the larvae are rarely found and this test is not usually diagnostic.

Intestinal hookworm infections, including A. ceylanicum, are diagnosed by identifying the eggs in the feces, as for animals.

A. caninum infections are difficult to diagnose. The parasite cannot be detected by examination of the feces, as the single worms found in human hosts do not produce eggs. The presence of eosinophilia aids in diagnosis. Ulcerations in the ileum and colon, and occasionally hookworms, may be seen during colonoscopy. Serologic tests for hookworms, including ELISAs and immunoblotting (Western blotting) are used only in research.

Treatment

Intestinal Disease

Hookworms in the intestines can be treated with a variety of anthelmintics. Iron therapy or blood transfusions may also be needed. In hookworm endemic regions, reinfection is likely, in that currently utilized anti-nematode drugs do not confer lasting protection.

Cutaneous Larva Migrans

Cutaneous larva migrans can be treated with topical or oral anthelmintics. Patients with minimal symptoms may not require treatment, as the infection is self-limiting. Pruritus usually decreases within 24 to 48 hours after the initial treatment, and the lesions usually resolve in a week.

Prevention

Zoonotic hookworm disease is best prevented by eliminating the parasites from dogs and cats to decrease contamination of the soil. Dogs and cats should also be kept off beaches and other places where children play in the sand. Sandboxes should be covered when not in use. Sodium borate can be used to sterilize lawns, kennels or other areas. Removing canine feces at least twice a week also decreases soil contamination. The larvae do not survive well in dry, bare soil in direct sunlight.

Prolonged skin contact with contaminated soil should be avoided. Wearing footwear (or gloves when gardening) can decrease the risk of infection. A waterproof sheet may be spread over damp work areas, when working under houses or in other potentially contaminated areas. When visiting the beach, wearing protective footwear, and using waterproof groundcovers rather than towels for sunbathing can help prevent contact with potentially contaminated sand.

Hookworms can be acquired by ingestion as well as skin contact; therefore, unsafe water should be boiled, and potentially contaminated foods avoided. The hands should also be washed before eating and after contact with soil or other sources of hookworms.

Morbidity and Mortality

Cutaneous larva migrans is particularly common in children. The prevalence is also high in adults who have close contact with soil or sand, including gardeners, farmers, miners, exterminators and others whose occupations require crawling under houses. Cutaneous larva migrans is the most common travel-related skin infection in tourists to tropical areas. Infections are self-limiting and the lesions resolve within a few weeks to a few months. Rare cases have lasted up to a year. Deaths are not seen.

Intestinal infections with A. caninum seem to be uncommon. Human intestinal infections with A. ceylanicum are common in some parts of Asia, and possibly other regions. Classic intestinal hookworm disease caused by human hookworms can result in anemia, malnutrition, and, in severe cases, congestive heart failure, severe blood loss or death, especially in young children. The potential for zoonotic intestinal hookworms (A. ceylanicum) to cause severe disease (including non-blood loss related cognitive impairment) or death following natural infection needs further study.

Ancylostoma Duodenale

Ancylostoma duodenale, the Old World hook worm is a very common nematode parasite in the small intestine of man. It causes "ancylostomiasis" in man, and until recently this hookworm ranked as the most important helminthic infection of man, but it has been brought under control in many countries.

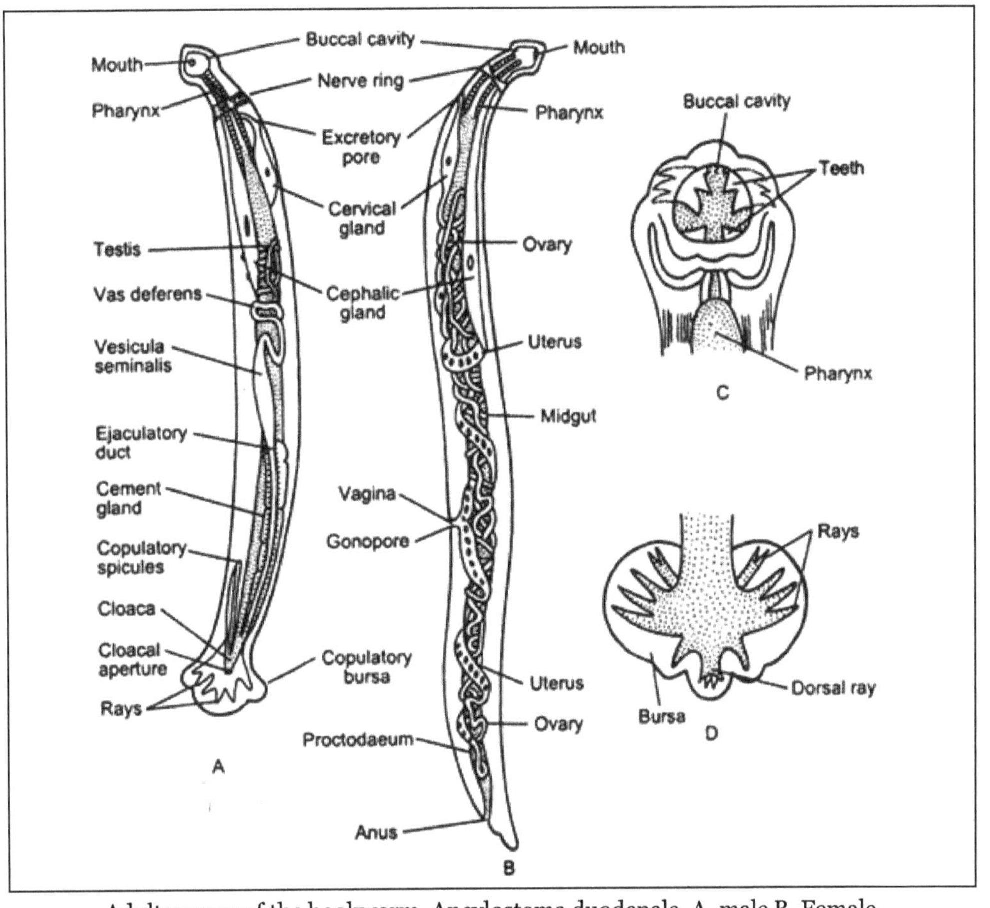

Adult worms of the hookworm, Ancylostoma duodenale. A. male B. Female.
C. Mouth region of Ancyiostoma, D. Posterior and of a male.

A. Dubini, an Italian physician first discovered the parasite during the autopsy of a woman in 1838. Perroncito (1881) described the development of free-living larvae in the soil. Looss described the pathogenesis and mode of entrance of the larvae in the intestine of the man in 1898.

Geographical Distribution of Ancylostoma Duodenale

The infection of the parasite has been reported among the rural people of the tropical countries and it may also occur in temperate countries where the temperature and humidity are favourable for the development of the larvae in the soil.

Incidence of the hookworm has been reported from Europe, Egypt, India, Bangladesh, Sri Lanka, Central and North China, and Pacific Islands. In India it is more predominant in Punjab, Haryana, and Himachal Pradesh.

Habitat of Ancylostoma Duodenale

The adult hookworms reside in the small intestine-of man particularly in the jejunum, less often in the duodenum, and rarely in the ileum. Thus, the name of the parasite appears to be a misnomer. It has been found on rare occasions in pigs. The adult worms anchor the wall of the small intestine by their anterior ends.

Morphology of Ancylostoma Duodenale

Adult Worms

1. The mature worms are cylindrical in shape, plump, rigid and creamy-white in colour.
2. The anterior end is bent dorsally like a hook (hence the name is "hookworm") and provided with the dorsally placed oral aperture.
3. The oral aperture is provided with 6 sharp teeth (cutting plates) on the ventral aspect, two teeth are found on each side, and two on the dorsal surface.
4. The buccal capsule is large and conspicuous, and is lined with chitin-like substance.
5. The cuticle is with fine transverse striations.
6. A minute finger-like cervical papillae is present on each side, a little away from the anterior extremity.
7. Two cephalic glands, a small oesophageal gland and two pear-shaped cervical glands are connected to the oesophagus, the secretion of the oesophageal gland prevents clotting of the ingested blood.

But Thorson (1956) reported that the oesophageal gland opens near the cutting plates (teeth) in the buccal capsule and participate in the extracorporeal digestion. The function of cervical glands and cephalic glands is not known.

1. Sexual dimorphism is distinct. The female is slightly larger and has a straight and pointed candal end The male is characterized by bursa copulatrix (an invagination of the body-wall around the genital aperture) at its caudal end. The bursa is supported by 13 rays. It has two protrusible spicules (1 mm in length) which assist in sperm transfer during the copulation.
2. The female worms measure about 10 to 13 mm x 0.6 mm while the males measure about 8 to 10 mm x 0.5 mm.
3. The female gonopore is separate and is located at the junction of the posterior and in the middle third.
4. The male has a cloaca where the ejaculatory duct opens.

Eggs

1. The eggs are oval and colourless with broadly rounded extremities, and measure about 60 μm x 40 μm.
2. Each egg has a thin outer shell and a very fine vitelline layer.
3. When freshly passed the egg has the segmented ovum surrounded by a clear space. It is usually segmented into 2 to 8 cells.

Life Cycle of Ancylostoma Duodenale

The life cycle of Ancylostoma duodenale is completed within a single host (man), hence it is called monogenetic. No intermediate host is recorded in the life cycle of A. duodenale.

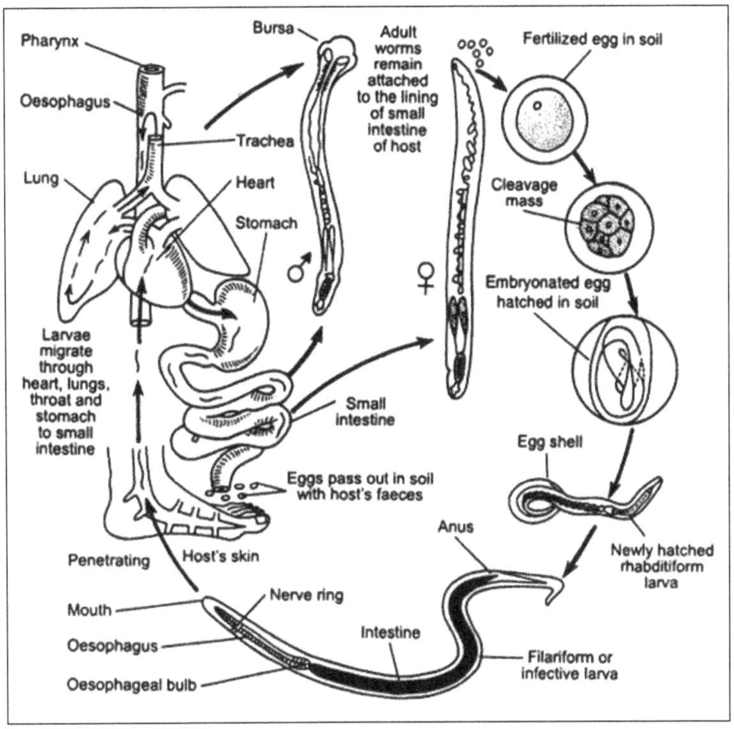

Life cycle of hookworm, Ancylostoma duodenale.

Copulation and Fertilization

The male and female worms mate in the intestine of man. During copulation, the bursa of male is applied on the vulva of female and sperms are transferred in the seminal receptacles where fertilization takes place.

Passage of the Eggs from the Infected Host

A single female worm lays about 10 to 20 thousand eggs per day. Fertilized eggs come out of the hosts body alone with the faeces. The contained ovum undergoes segmentation during the passage through the bowel. The eggs are not infective to man but the larvae hatched out of the eggs in soil represents the infective stage.

The larva hatches out of the egg in 24 to 48 hours in damp soil with warmth (27 °C) and sufficient oxygen. It is to be mentioned that very little development takes place in diluted faeces due to its high acidic pH (4.8-5).

Development of Larva in Soil

The newly hatched larva is called rhabditiform larva or first stage juvenile owing to its rhabditic oesophagus. It measures about 250 μm in length and possesses a rounded anterior end and a sharp caudal end.

It is very active and feeds voraciously on the organic matters present in the stool and soil. It grows rapidly and moults twice in one week: the first on the 3rd day and the second on the 5th day. By this time the larva attains a length of 500 to 600 μm.

After the second moult, the oesophageal bulb disappears and the organ becomes simple and muscular. The shedded cuticle is retained as the protective sheath. It is now called a filariform larva or third stage juvenile which is infective to man. Although actively motile, the larva ceases to feed and does not grow further. Development from egg to filariform larva takes about 8-10 days.

Biology of the Larvae

The larvae live in the upper 1/2 inch of the soil. In tropical climates, the life span is short; 90% of them die within 2 to 3 weeks, and the rest by 6 weeks. Filariform larvae are readily destroyed by direct sunlight. They are very sensitive to desiccation and extremes of temperature. They also move towards the source of oxygen.

Entrance into a New Host

The filariform larvae gain entrance into the body of the host (man) by penetrating the skin. They cast off their sheaths and bore through the skin directly or through the hair- follicles.

The warmth of sole of legs and hands is believed to attract the filariform larvae and hence the infection is more common to the bare-footed subjects. Infection may also happen from handling faeces-soiled clothing's if they are left damp for 4 or 5 days. There has been reported that intra-uterine infections are also possible.

Migration after Entrance

After several hours of infection, the larvae enter into lymphatics or small venules. The larvae unable to reach the vascular spaces usually die or are phagocytosed. They are passively carried to the right part of the heart and then to the lungs by way of pulmonary artery, on about 3rd day.

They then migrate in the alveoli by boring the wall of pulmonary capillaries where they are arrested owing to their large size. From the lungs the larvae ascend the bronchi, trachea, epiglottis to the back of the pharynx and are swallowed.

During the migration when they reach the oesophagus, the third moulting occurs with the formation of a temporary buccal capsule bearing 4 small teeth. Finally, they reach the jejunum via the stomach and the duodenum between the 7th and the 10th day. By this time, the larva grows rapidly and attains a size of 2 mm × 0.13 mm.

Establishment and Laying of Eggs

The growing larva settles in the jejunum and undergoes the 4th moult and final moult on the 15th day. During this moult the definitive buccal capsule is formed in place of temporary buccal capsule with teeth. The worm becomes sexually mature in 3 to 4 weeks and the eggs begin to appear in the stool on the 6th week.

The wounds are afflicted by the sharp teeth of adult worms in the intestinal wall at the region of attachment. Through these wounds blood comes out and the worms then suck the blood by the action of suctorial pharynx.

Coagulation of the blood is prevented by the production of a secretion from the mouth which has an anticoagulatory property and thus blood is prevented from being coagulated. A. duodenale is recorded to suck profuse quantity of blood and the rate of sucking of blood is recorded to be 0.8 ml in 24 hours.

Pathogenicity of Ancylostoma Duodenale

The infective larvae cause some kinds of eruption of the skin which may be pustular, known as ground-itch. The hosts infected with adult worms become anaemic as a result of considerable loss of blood from the intestinal wounds.

Besides the anaemic condition, the hosts become susceptible to all diseases, particularly tuberculosis because of the damage done by the larvae during their transit from the heart to the final abode.

Treatment and Control of Disease caused by Ancylostoma Duodenale

The hookworm infection can be checked by administering tetrachlorethyl, hexylresorcinol, carbon tetrachloride and blephenium, etc.

Control Measures

1. The most important is the sanitary disposal of human faeces to prevent pollution of the earth.
2. Borehole latrine should be introduced in the villages to reduce the spread of larva.
3. Boots and gloves should be weared while working in the garden.

Enterobius Vermicularis

Popularly known as pinworm and infects children. The adult worms are very small in size and inhabit the intestine, caecum and appendix. Sexes are separate. The gravid females migrate to the rectum and produce itching sensation. Eggs are laid in the rectum and infection occurs by bodily transference of eggs.

Dracunculus Medinensis

Commonly known as Guinea-worm and man it its final host. Young ones occur in a shrimp-like form of crustacea called Cyclops. The cyclops are ingested along with drinking water and the young parasites are liberated into the stomach from where they migrate to subcutaneous tissues.

Males measure about 4 cm and die after copulation. The females are about 12 cm in length and gravid females have two uteri full of fertilized eggs. For laying of eggs the females come to the superficial layer of the skin and die after discharging eggs.

Necator Americanus

Classification

The Phylum is Nematoda and its common name is “Hookworm”. Its literal meaning is “American Murderer” because of the large amounts of blood it robs from the human host’s body.

Morphology of Necator Americanus

The male adult measures 7 to 9 mm long, while the female is slightly larger measuring 9 to 11 mm. The anterior of the worm curves in the opposite direction of the body curvature resulting in a hooked appearance. The mouthparts of hookworms have buccal capsules with cutting plates on the ventral surface and small teeth inside the capsule on the dorsal surface.

The posterior end of the male has a copulatory bursa, which is used to catch and hold the female during mating. The females have a vulva near the center of the body, slightly anterior.

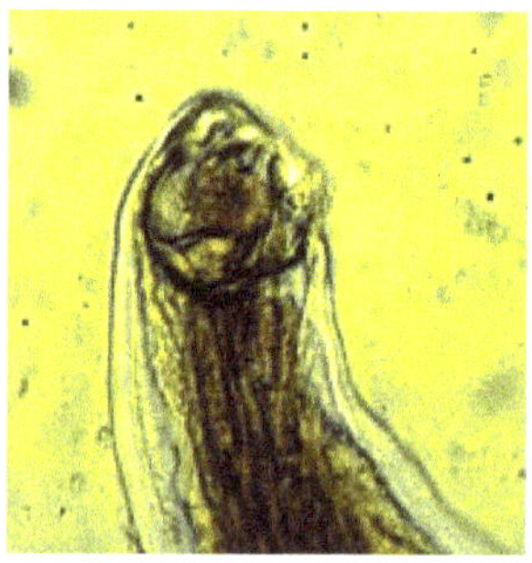

Mouth with cutting plates.

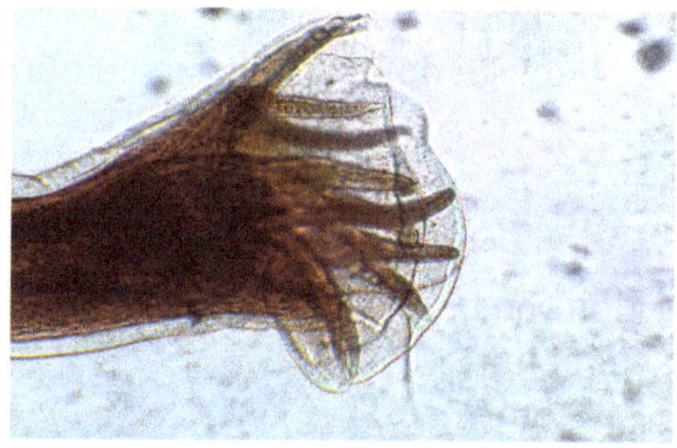

Copulatory bursa of male.

The eggs of Necator americanus range 57 – 76 µm by 35 – 47 µm. They are oval or ellipsoidal with a thin shell. The eggs cannot be distinguished from the eggs of Ancylostoma duodenale.

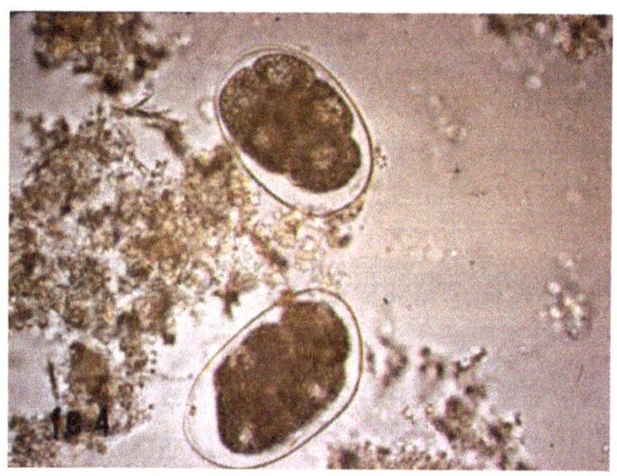

Egg in stool.

Lifecycle and Epidemiology

The adult Necator americanus worm can be found in the lumen of human small intestines. The eggs are passed in the feces and usually hatch in 1 to 2 days. The eggs need a shady, moist, and warm environment in order to survive. The eggs release the rhabdidiform larvae or J_1 which can grow in feces or soil. After 5 to 10 days, the J_1 will molt twice and become the filariform or J_3 larvae. This is the infective stage of the worm and it can survive for 3 to 4 weeks in favorable environmental conditions. Once the worm makes contact with a potential human host, the J_3 can penetrate the skin. It then migrates through the veins to the heart and then lungs. The J3 then penetrates into the alveoli and ascend into the throat until swallowed. The J_3 will continue on until it reaches the small intestine, where it attaches to the intestinal wall and matures into an adult worm. The usual life span for the adult worm is from 1 to 2 years.

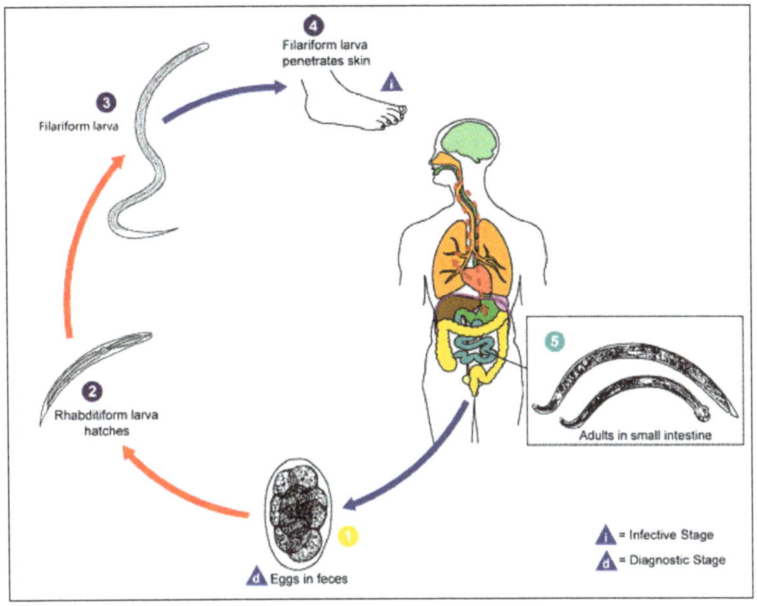

Geographic Distribution

The Necator americanus is the second most common human helmith infection. It infects over 1 billion people worldwide. It thrives in warm, moist climates and is said to predominate in the Americas and Australia. As can be seen in the map, it basically infects much of the area called the tropics. This is the area between the Tropic of Cancer and the Tropic of Capricorn.

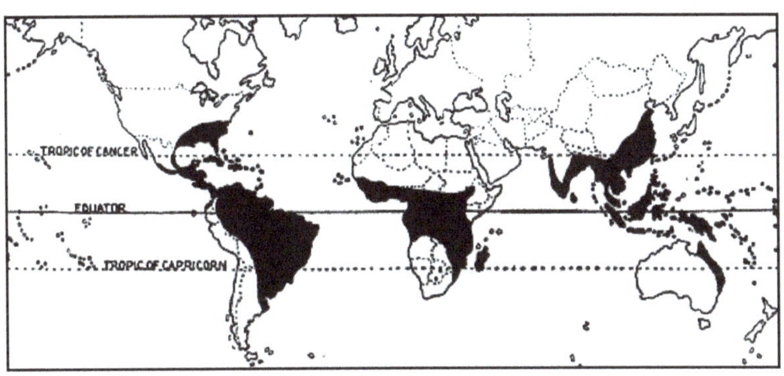

Geographic distribution of Necator americanus.

Pathology and Symptoms

The pathology of Nector americanus can be divided into two areas: larvae and adult worms. The larvae penetrate the skin and travel through the lymph nodes. Once in the lymph nodes, the larvae migrate between the bloodstream, lungs, and small intestine. The most common symptoms of this infective stage are itching due to skin penetrations and respiratory problems during the pulmonary migration of the larvae. The adult worms attach to the villi of the small intestines. The villi are tiny, fingerlike projections on the surface of the small intestine that help absorb nutrients. Once attached to the wall of the small intestinal, the hookworm's mouthparts penetrate blood vessels and suck blood. One Necator americanus will take approximately 30 µl of blood daily. Light affections appear to be asymptomatic. Heavier infections include abdominal discomfort, diarrhea, cramps, anorexia, and weight loss. The loss of nutrition in the blood can lead to iron deficiency and anemia. In more severe cases, iron deficiency anemia can cause physical and mental retardation in children. Death is rare in humans.

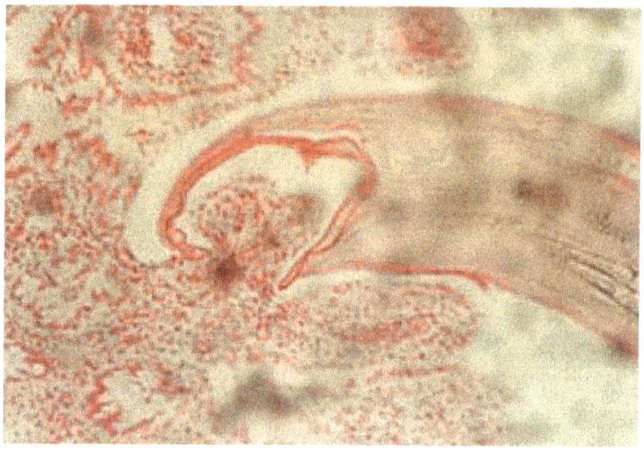

Adult attached to the small intestine.

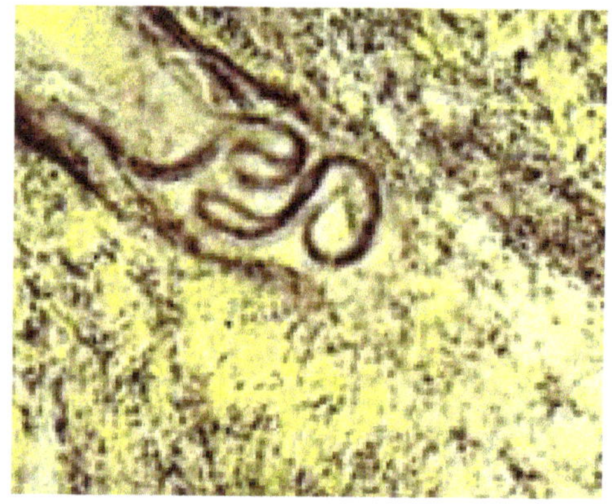

Migrating larvae in tissue.

Diagnosis

The most common method for diagnosing hookworm is through identification of eggs in a fecal sample.

Treatment

The chosen drug for treating Necator americanus is Albendazole. A study of Albendazole in treating Necator americanus in Sri Lanka has shown a 77.9% cure rate and a 94.5% reduction in egg count. An alternative drug is Mebendazole which has shown a 95% cure rate and a 99.9% reduction in egg count. However, Mebendazole is more costly and less convenient than Albendazole. Iron supplements should be included in the treatment of individuals with Necator americanus.

Public Health Strategies and Eradication

Eradication is difficult or impossible due to the nature of Necator americanus. Like other worms we have studied, N. americanus locates itself in the human body where it is not easily controlled by the human immune system. In addition, poorer countries do not have the funds to improve sewage treatments. Therefore, it is important to educate a community on the importance of proper sanitation and investing in footwear. Some communities have invested in community-wide albendazole chemotherapy every 18 months.

Ancylostoma Caninum

Ancylostoma caninum is a species of nematode known as a hookworm, which principally infects the small intestine of dogs. The result of A. caninum infection ranges from asymptomatic cases to death of the dog; better nourishment, increasing age, prior A. caninum exposure, or vaccination are all linked to improved survival. Other hosts include carnivores such as wolves, foxes, and cats, with a small number of cases having been reported in humans.

Warm and moist conditions are important to allow survival of A. caninum during the free-living stages of its lifecycle, so it is largely restricted to temperate, tropical, and subtropical regions. In parts of the world where these climatic requirements are met such as Sri Lanka, Southeast Asia, and Malaysia, A. caninum is the main cause of hookworm disease in canines.

Morphology

A. caninum females are typically 14–16 mm (0.55–0.63 in) long and 0.5 mm (0.02 in) wide, while the males are smaller at 10–12 mm (0.39–0.47 in) in length and 0.36 mm (0.01 in) in width. On males, a copulatory bursa exists, which during copulation, attaches the female via a roughly 0.9-mm-long, spine-like spicules positioned on three muscular rays. As with other nematodes, the sperm lack flagella. The copulatory bursa is a unique feature of Strongylida members, thus making it a useful means for identifying members of this suborder; it is also used to distinguish members within the suborder due to differences in bursa appearance between species. The vulva of A. caninum females is located at the boundary of the second and final thirds of the body.

The teeth of A. caninum are found in the buccal capsule and divided into three sets. Two ventral sets form a lower-jaw equivalent, while a further set projects from the dorsal side and loosely equates to an upper jaw. Each ventral set has three points, with those furthest to the sides being the largest. While the ventral sets are prominent, the dorsal set is hidden deeper in the buccal capsule.

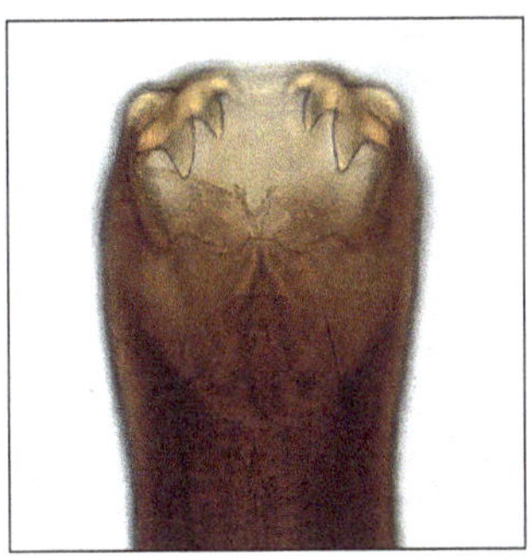

Mouth and teeth of A. caninum.

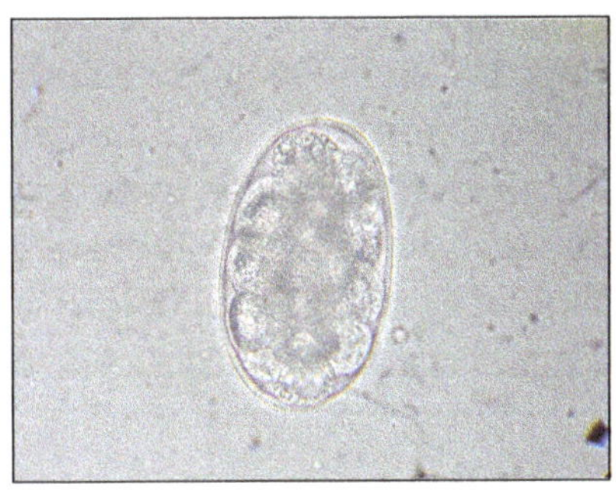

A. caninum egg.

A. caninum bends its head end upward (dorsally), which has in the past been noted as a potential source of confusion when determining how the hookworm is oriented. If it has recently ingested blood, A. caninum is red in colour; if not, it appears grey. A. caninum has an alimentary canal made up of an esophagus, intestine, and rectum – the esophagus is highly muscular, reflecting its role in pulling intestinal mucosa into the body when it feeds. Esophageal and anal rings of A. caninum are the source of nerve fibres that extend throughout the body to innervate sensory organs including amphids and phasmids.

Eggs are laid by the females typically when at the eight-cell stage. Eggs are 38–43 μm in width with thin walls.

Distribution

Freezing, heating above 37 °C (99 °F), drying or exposing A. caninum to sunlight all give reduced survival of the free-living stage, with rates of infection rising with temperature provided 37 °C is not exceeded. A. caninum is, therefore, largely restricted to warm, moist climates, though infections are seen in the United States and southern Canada where the temperature is suboptimal. Specific niches are also able to satisfy the environmental requirements of A. caninum, despite not necessarily being in the tropics, such as mines.

Lifecycle

Transmission via the Environment

Eggs are excreted from host in the feces and typically hatch within a day on moist, warm soil giving larvae with a non-living cuticle layer. By 4–5 days, the larvae have moulted twice and are now able to infect a host. Migration occurs from the faeces into the surrounding soil. Two routes of infection from the environment exist. The first route involves penetration of skin at hair follicles or sweat glands, especially between the footpads where contact with soil is frequent and the skin is thinner than otherwise. Secretion of a protease by A. caninum is thought to aid this process. The larvae then migrate through the dermis of the skin, enter the circulatory system and are carried to the lungs. A. caninum larvae exit the blood at the lungs, move from the alveoli up through the trachea and are swallowed to end up in the intestine.

The second and more common route to the small intestine is by direct ingestion of A. caninum by the host, but the subsequent process is identical in either case. During this third stage of the larva, male or female reproductive organs become established. Larvae of this stage have been shown to secrete a molecule (Ac-asp-2) related to venom allergens in response to host-specific signals; this is thought to have a possible role in helping with the infection process. A third and final moulting occurs to give the mature form of A. caninum which then feeds on mucosa and blood of the small intestine wall. The trigger of feeding is understood to be a receptor-mediated response, however the detail of this process has yet to be established. Sexual reproduction also occurs in the intestine to give a further round of eggs to complete the cycle. Females are thought to produce a pheromone which attracts males and are able to produce about 10,000 eggs per day.

Direct Transmission

Direct transmission between hosts is also possible. Larvae having accessed through the skin may avoid exit via the lungs and remain in circulation for transport around the body. At the uterine artery of a pregnant female, the larvae are able to cross the placenta to cause prenatal infection of foetuses. Larvae of an infected foetus move to the liver until birth, when migration continues with movement to the intestine via the circulation and lungs as previously described. Alternatively, A. caninum larvae evading exit from the circulation at the lungs may instead be carried to the mammary glands and transmitted from the mother in her colostrum or milk to her pups; infection then proceeds in an identical manner to infection by ingestion from the environment. Infected mothers have been found to only rarely give prenatal transmission to pups, while the likelihood of nursing of pups causing transmission (via the lactational route) is much higher.

Pathogenesis

Damage During Migration to Intestine

A. caninum larvae cause damage to the host at the point of entry through the skin, leaving a wound vulnerable to secondary infections. As the larvae migrate through the skin, an inflammatory response, dermatitis, is often stimulated, which can be exacerbated in hosts which have hypersensitive responses. Further damage is caused when the larvae leave the circulation and enter the lung, with the amount of damage dependent on the extent of the infection; pneumonia and coughing are common consequences.

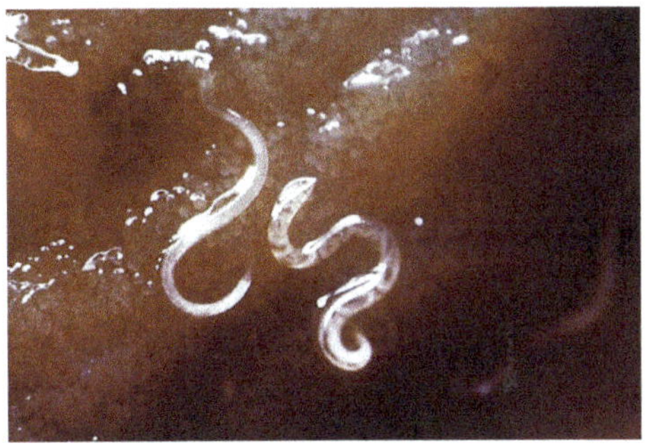

Pair of A. caninum hookworms.

Damage Once in Intestine

Once in the gut, A. caninum attaches to and ingests the mucosal lining along with some consumption of blood; up to 0.1 ml in 24hrs. In a 24hr period A. caninum typically feeds from six sites. This damage to the mucosa compromises the body's defences and can result in secondary infections by microbes. A group of anticoagulant proteins called A. caninum anticoagulant proteins (AcAPs), which inhibit a range of blood coagulation factors such as Xa, are used by A. caninum to help in the feeding process by preventing clotting and increasing blood loss. These AcAPs are among the most powerful natural anticoagulants that exist and are a key reason for anemia being caused and blood being observed in the faeces of infected hosts. Blood losses peak just prior to egg production by the females because this is when their requirements for food are greatest; the amount that they are eating is also peaking, so maximal damage to the intestine is being caused.

Diagnosis

Analysis of faeces is the definitive method by which a suspected A. caninum infection is confirmed. The faeces are sampled and the characteristic ovular, thin-shelled eggs of A. caninum looked for. Absence of eggs in faeces does not rule out infection; a significant delay of at least 5 weeks exists between initial infection and excretion of eggs in the faeces (larvae must fully mature and reproduce before eggs can be laid). In fact, pups frequently die before passing of eggs in the faeces begins. Using the number of eggs in stool samples as an indicator of the extent of infestation requires care to be taken because females have been shown to produce fewer eggs each when the overall number of worms increases.

Signs and symptoms expected to be observed together with A. caninum eggs in the faeces are lethargy, weight loss, weakness, roughness of the hair coat, and pale mucous membranes indicative of anemia. Well-fed, older dogs with smaller infestations may present few or even none of these symptoms. Diarrhoea is rare, but stools are typically black due to the blood-derived haemoglobin present in them.

The disease resulting from such A. caninum infection is referred to by the general term hookworm disease or the more specific ancylostomiasis and ancylostomosis, which recognise the genus of the causative nematode.

Prevention and Control

A clean environment minimises the risk of A. caninum infection; this can include regularly washed concrete or gravel in kennels instead of soil. Females are typically checked prior to using them for breeding purposes for nematodes such as A. caninum and birth and suckling can be restricted to sanitised areas to lower the risk of health complications to the pups. When infection of a pregnant dog is known or suspected fenbendazole or ivermectin can be administered to the female to help avoid transmission to the pups.

Canines have been seen to develop significant resistance to A. caninum naturally with age; this protection develops faster in and fully mature females show substantially greater resistance than fully mature dogs. Specifically the age-related resistance means A. caninum takes longer to reach sexual maturity in older animals and fewer larvae fully develop.

Dog kennel: keeping kennels clean reduces risk of A. caninum infection.

Numerous vaccines have been developed with varying success against A. caninum. Use of an enzyme important in the worm's feeding process is popular with one example being AcCP2, a protease, which when used to vaccinate dogs gives a strong antibody response, lowering of numbers of eggs found in stools and a decrease in intestinal worm size. These effects are attributed to reduced AcCP2 activity upon antibody binding. A similar approach has been taken using another A. caninum digestive enzyme, AcGST1, but it failed to give statistically significant results in dogs.

An alternative approach has been to disrupt the migratory ability of A. caninum, this was done so successfully using the AcASP1 protein of A. caninum which gives increases in antibody levels of all subclasses and a reduced worm burden. Other studies using the same vaccine have shown statistically significant 79% reductions in worm burden resulting from this approach.

Animals with prior exposure to A. caninum show enhanced resistance but careful removal of all worms from the previous infection results in loss of this improvement. Studies in mice show resistance due to past exposure can protect against otherwise lethal worm doses and that this is a general form of resistance – defence is offered against subsequent infections via either mouth or skin.

Medication

Drugs used in treatment of A. caninum infections of dogs include: dichlorvos, fenbendazole, flubendazole, mebendazole, nitroscanate, piperazine, pyrantel, milbemycin, moxidectin, diethylcarbamazine, oxibendazole, and ivermectin.

In Humans

In inappropriate hosts such as humans, A. caninum is able to enter the skin, but cannot proceed into the circulation and on to the intestine; instead, the disease dermal larva migrans results, caused by movement of the nematode within the skin and which can persist for several months without intervention.

While access to the intestine is not possible via this route, it can occur via ingestion; in a report of 93 enteritis cases in northern Queensland, Australia, which were possibly caused by A. canium infection, all those interviewed described behaviour consistent with A. caninum exposure and a colonoscopy of one patient gave positive identification of an adult A. caninum worm. Since then, work has shown A. caninum can easily go unnoticed or fail to be preserved in specimens, making the true incidence of infection in humans likely to be higher than is officially recorded.

Threadworms

Threadworms, also known as pinworms, are tiny parasitic worms that infect the large intestine of humans. Threadworms are a common type of worm infection in the UK, particularly in children under the age of 10. The worms are white and look like small pieces of thread. You may notice them around your child's bottom or in their poo. They don't always cause symptoms, but people often experience itchiness around their bottom or vagina. It can be worse at night and disturb sleep.

Pinworm infection is the most common type of intestinal worm infection in the United States and one of the most common worldwide. Pinworms are thin and white, measuring about 1/4 to 1/2 inch (about 6 to 13 millimeters) in length.

While the infected person sleeps, female pinworms lay thousands of eggs in the folds of skin surrounding the anus. Most people infected with pinworms have no symptoms, but some people experience anal itching and restless sleep.

Pinworm infection occurs most often in school-age children, and the microscopic eggs are easily spread from child to child. Treatment involves oral drugs that kill the pinworms and thorough washing of pajamas, bed linens and underwear. For best results, the entire family should be treated.

Symptoms

Symptoms of pinworm infection may include:

- Itching of the anal or vaginal area;
- Insomnia, irritability and restlessness;
- Intermittent abdominal pain and nausea.

When to see a Doctor

Consult your doctor if you have severe anal itching, especially at night.

Causes

Accidentally swallowing or breathing in pinworm eggs causes a pinworm infection. The microscopic eggs can be carried to your mouth by contaminated food, drink or your fingers. Once swallowed, the eggs hatch in the intestines and mature into adult worms within a few weeks.

Female pinworms move to the anal area to lay their eggs, which often results in anal itching. When you scratch the itchy area, the eggs cling to your fingers and get under your fingernails. The eggs then get transferred to other surfaces, such as toys, bed linens or toilet seats. The eggs can also be transferred from contaminated fingers to food, liquids, clothes or other people.

Pinworm eggs can survive for two to three weeks on surfaces.

Risk Factors

Risk factors for pinworm infection include:

- Being young: Pinworm infections are most likely to occur in children ages 5 to 10. The microscopic eggs are easily spread to family members, caregivers, or other children at school or child care centers. Pinworm infections are uncommon in children younger than age 2.
- Living in crowded spaces: People who live in institutions are at higher risk of developing pinworm infections.

Complications

Typical pinworm infections don't cause serious problems. In rare circumstances, heavy infestations can cause infection of female genitals.

The parasite can travel from the anal area up the vagina to the uterus, fallopian tubes and around the pelvic organs. This can cause problems such as inflammation of the vagina (vaginitis) and inflammation of the inner lining of the uterus (endometritis).

Although rare, other complications of a pinworm infection may include:

- Urinary tract infections;
- Weight loss.

Prevention

Pinworm eggs can cling to surfaces, including toys, faucets, bedding and toilet seats, for two weeks. So besides regular cleaning of surfaces, methods to help prevent the spread of pinworm eggs or to prevent reinfection include:

- Wash in the morning: Because pinworms lay their eggs at night, washing the anal area in the morning can help reduce the number of pinworm eggs on your body. Showering may help avoid possible recontamination in bath water.

- Change underwear and bed linen daily: This helps remove eggs.
- Launder in hot water: Wash bedsheets, nightclothes, underwear, washcloths and towels in hot water to help kill pinworm eggs. Dry on high heat.
- Don't scratch: Avoid scratching the anal area Trim your child's fingernails so there's less space for eggs to collect. Discourage nail biting.
- Wash your hands: To reduce your risk of getting or spreading an infection, wash your hands thoroughly after having a bowel movement or changing a diaper and before eating.

Human Whipworm

Trichuris trichiura, also known as the human whipworm, is a roundworm that causes trichuriasis in humans. It is referred to as the whipworm because it looks like a whip with wide handles at the posterior end. The whipworm has a narrow anterior esophagus and a thick posterior anus. The worms are usually pink and attach to the host via the slender anterior end. The size of these worms varies from 3 to 5 cm. The female usually larger than the male.

A whipworm infection, also known as trichuriasis, is an infection of the large intestine caused by a parasite called Trichuris trichiura. This parasite is commonly known as a "whipworm" because it resembles a whip.

A whipworm infection can develop after ingesting water or dirt contaminated with feces containing whipworm parasites. Anyone who has come into contact with contaminated feces can also contract a whipworm infection. The infection most often occurs in children. It is also more common in people who live in regions with hot, humid climates and in areas with poor hygiene and sanitation.

Approximately 600 to 800 million people around the world have a whipworm infection. This type of infection can also occur in animals, including cats and dogs.

Symptoms of a Whipworm Infection

A whipworm infection can cause a variety of symptoms, ranging from mild to severe. They may include the following:

- Bloody diarrhea,
- Painful or frequent defecation,
- Abdominal pain,
- Nausea,
- Vomiting,

- Headaches,
- Sudden and unexpected weight loss,
- Fecal incontinence, or the inability to control defecation.

Causes a Whipworm Infection

A whipworm infection is caused by a parasite called Trichuris trichiura. This parasite is also known as a "whipworm" because it is shaped like a whip. It has a thick section on one end that resembles the whip handle, and a narrow section on the other end that looks like the whip.

People typically get whipworm infections after consuming dirt or water contaminated with feces containing whipworm parasites or their eggs. Whipworm eggs can get into the soil when contaminated feces are used in fertilizers or when an infected person or animal defecates outside.

Someone might unknowingly ingest the whipworm parasites or their eggs when they:

- Touch the dirt and then put their hands or fingers in or near their mouth.
- Eat fruits or vegetables that haven't been thoroughly washed, cooked, or peeled.

Once they reach the small intestine, whipworm eggs hatch and release larvae. When the larvae mature, the adult worms live in the large intestine. The female worms usually begin to deposit eggs about two months later. According to the Centers for Disease Control and Prevention, the females shed between 3,000 and 20,000 eggs per day.

Risk Factors for a Whipworm Infection

A whipworm infection can occur in anyone. However, people may be more likely to contract a whipworm infection if they:

- Live in a region with a hot, humid climate.
- Live In An area with poor sanitation and hygiene practices.
- Work In an industry where they come into contact with soil that contains manure.
- Eat Raw vegetables that are grown in soil fertilized with manure.

Children also have a higher risk of getting a whipworm infection. They often play outdoors and might not wash their hands thoroughly before eating.

Whipworm Infection Diagnoses

To diagnose a whipworm infection, your doctor will order a stool test. You will be required to give a sample of your feces to a lab for testing. The stool test can determine whether there are whipworms or whipworm eggs in your intestines and feces.

This type of test shouldn't cause any discomfort or pain. Your doctor will give you a sterile container and a kit containing plastic wrap and special bathroom tissue. Place the plastic wrap loosely over the toilet bowl and make sure it's held in place by the toilet seat. After you have a bowel movement, use the special tissue to put the stool into the container. For infants, the diaper can be lined with the plastic wrap to collect the sample. Make sure to wash your hands thoroughly after the test.

The sample will be sent to a lab, where it will be analyzed under a microscope for the presence of whipworms and their eggs.

Whipworm Infection Treatment

The most common and effective treatment for a whipworm infection is an antiparasitic medication, such as albendazole and mebendazole. This type of medication gets rid of any whipworms and whipworm eggs in the body. The medication usually needs to be taken for one to three days. Side effects are minimal.

Once your symptoms subside, your doctor may want to perform another stool test to make sure the infection is gone.

Outlook for Someone with a Whipworm Infection

Most people who receive treatment for a whipworm infection make a full recovery. When left untreated, however, the infection can become severe and cause complications. These include:

- Delayed growth or cognitive development.
- Infections in the colon and appendix.
- Rectal prolapse, which occurs when a section of the large intestine protrudes from the anus.
- Anemia, which occurs when the number of healthy red blood cells drops too low.

Prevention of Whipworm Infection

To reduce your risk of contracting a whipworm infection, you should:

- Wash your hands thoroughly, especially before handling food.
- Wash, peel, or cook foods thoroughly before eating them.
- Teach children not to eat soil and to wash their hands after playing outdoors.
- Boil or purify drinking water that may be contaminated.
- Avoid contact with soil contaminated with fecal matter.
- Use caution around animal feces and clean up fecal matter when possible.
- Confine livestock, such as pigs, into pens. These enclosures should be thoroughly cleaned on a regular basis.
- Keep the grass cut short in areas where dogs or cats regularly defecate.

The spread of whipworm can be prevented in high-risk areas by installing effective sewage disposal systems.

Filarial Worms

Filarial worm is any of a group of parasitic worms of the family Filariidae (phylum Nematoda) that usually require two hosts, an arthropod (the intermediate host) and a vertebrate (the primary host), to complete the life cycle. The larval phase occurs within the body of a biting insect. The mature (reproductive) phase occurs in the body of an animal bitten by the insect.

The female worm produces large numbers of microscopic, active embryos called microfilariae that pass into the bloodstream of the primary host. The microfilariae may then enter the body of an insect as the insect bites the infected animal. The microfilariae grow into larvae in the insect's muscles and may then be passed to the primary host when the insect bites an animal. The larvae reach adulthood within the vertebrate host, and the cycle repeats. In mammals filarial worms cause a group of infectious disorders including heartworm, elephantiasis, and river blindness. These disorders are known collectively as filariasis. At present, more than 200 million people are infected with filarial parasites.

Wuchereria Bancrofti

Wuchereria bancrofti is a human parasitic roundworm that is the major cause of lymphatic filariasis. It is one of the three parasitic worms, together with Brugia malayi and B. timori, that infect the lymphatic system to cause lymphatic filariasis. These filarial worms are spread by a variety of mosquito vector species. W. bancrofti is the most prevalent of the three and affects over 120 million people, primarily in Central Africa and the Nile delta, South and Central America, the tropical regions of Asia including southern China, and the Pacific islands. If left untreated, the infection can develop into a chronic disease called Lymphatic filariasis. In rare conditions, it also causes tropical eosinophilia, an asthmatic disease. No vaccine is commercially available, but high rates of cure have been achieved with various antifilarial regimens and lymphatic filariasis is the target of the WHO Global Program to Eliminate Lymphatic Filariasis with the aim to eradicate the disease as a public-health problem by 2020.

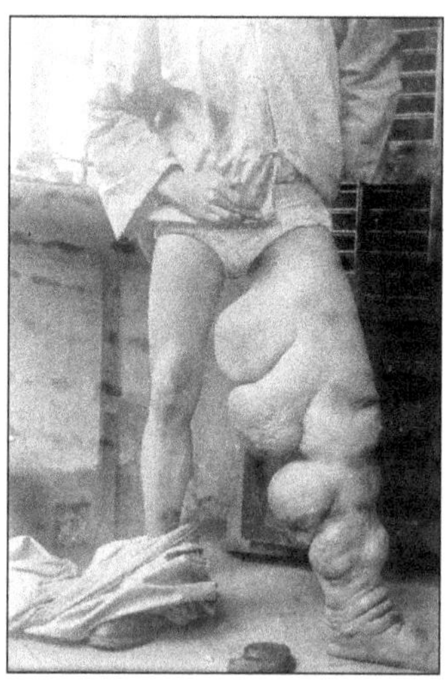

An infection in leg by W. bancrofti Specialty Infectious disease.

Signs and Symptoms

The pathogenesis of W. bancrofti infection is dependent on the immune system and inflammatory responses of the host. After infection, the worms mature within 6–8 months, male and female worms mate and then release the microfilariae. These microfilariae can be released for up to 10 years.

- The asymptomatic phase usually consists of high microfilaremia infection, and individuals show no symptoms of being infected. This occurs due to cytokine IL-4 suppressing the activity of TH1 cells in the immune system. This can occur for years until the inflammatory reaction rises again.
- In the inflammatory (acute) phase, the antigens from the female adult worms elicit inflammatory responses. The worms in the lymph channels disrupt the flow of the lymph, causing lymphedema. The individual exhibits fever, chills, skin infections, painful lymph nodes, and tender skin of the lymphedematous extremity. These symptoms often lessen after 5-7 days. Other symptoms that may occur include orchitis, an inflammation of the testes, which is accompanied by painful, immediate enlargement and epididymitis (inflammation of the spermatic cord).
- The obstructive (chronic) phase is marked by lymph varices, lymph scrotum, hydrocele, chyluria (lymph in urine), and elephantiasis. Microfilariae are not normally present in this phase. A key feature of this phase is scar formation from affected tissue areas. Other features include thickening of the skin and elephantiasis, which develops gradually with the attack of the lymphatic system. Elephantiasis affects men mainly in the legs, arms, and scrotum. In women, the legs, arms, and breasts are affected.

Parasite

Morphology

As a dioecious worm, W. bancrofti exhibits sexual dimorphism. The adult worm is long, cylindrical, slender, and smooth with rounded ends. It is white in colour and almost transparent. The body is quite delicate, making removing it from tissues difficult. It has a short cephalic or head region connected to the main body by a short neck, which appears as a constriction. Dark spots are dispersed nuclei throughout the body cavity, with no nuclei at the tail tip. Males and females can be differentiated by size and structure of their tail tips. The male worm is smaller, 40 mm (1.6 in) long and 100 μm (0.0039 in) wide, and features a ventrally curved tail. The tip of the tail has 15 pairs of minute caudal papillae, the sensory organs. The anal region is an elaborate structure consisting of 12 pairs of papillae, of which eight are in front and four are behind the anus. In contrast, the female is 60 millimetres (2.4 in) to 100 millimetres (3.9 in) long and 300 micrometres (0.012 in) wide, nearly three times larger in diameter than the male. Its tail gradually tapers and rounded at the tip. No additional sensory structures are seen. Its vulva lies towards the anterior region, about 0.25 mm from the head. Adult males and females are most often coiled together and are difficult to separate. Females are ovoviviparous and can produce thousands of juveniles known as microfilariae.

The microfilaria is a miniature adult, and retains the egg membrane as a sheath, and is often considered an advanced embryo. It measures 280 μm long and 25 μm wide. It appears quite structureless in vivo, but histological staining makes its primitive gut, nerve ring, and muscles apparent.

Lifecycle

W. bancrofti carries out its lifecycle in two hosts. Humans serve as the definitive host and mosquitos as the intermediate host. The adult parasites reside in the lymphatics of the human host. They are found mostly in the afferent lymphatic channels of the lymph glands in the lower part of the body. The first-stage larvae, known as microfilariae, are present in the circulation. The microfilariae have a membrane "sheath". This sheath, along with the area in which the worms reside, makes identification of the species of microfilariae in humans easier to determine. The microfilariae are found mainly in the peripheral blood and can be found at peak amounts from 10 pm to 4 am. They migrate between the deep and the peripheral, circulation exhibiting unique diurnal periodicity. During the day, they are present in the deep veins, and during the night, they migrate to the peripheral circulation. The cause of this periodicity remains unknown, but the advantages of the microfilariae being in the peripheral blood during these hours may ensure the vector, the night-time mosquito, will have a higher chance of transmitting them elsewhere. Physiological changes also are associated with sleeping, such as lowered body temperature, oxygen tension, and adrenal activity, and an increased carbon dioxide tension, among other physical alterations, any of which could be the signals for the rhythmic behavior of microfilarial parasites. If the hosts sleep by day and are awake at night, their periodicity is reversed. In the South Pacific, where W. bancrofti shows diurnal periodicity, it is known as periodic.

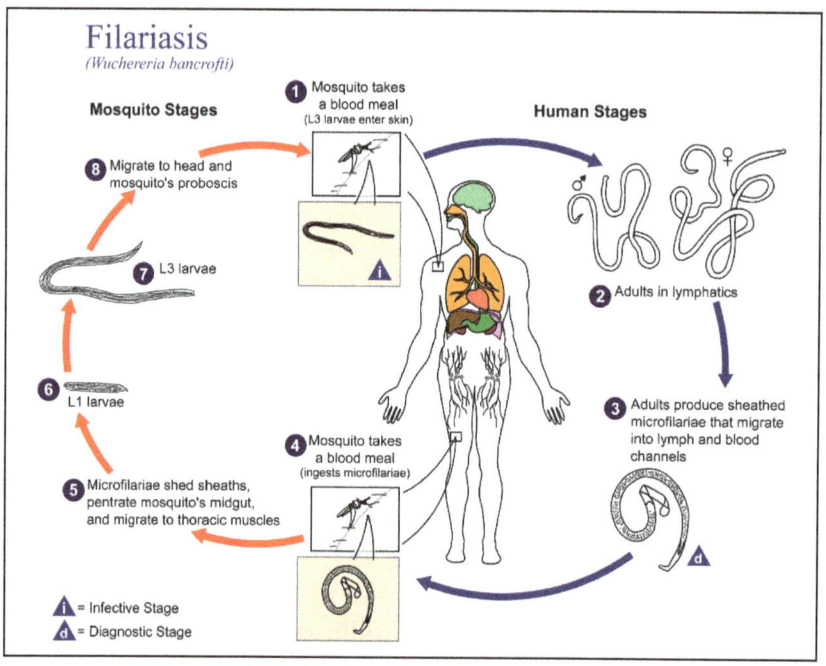

Lifecycle of Wuchereria bancrofti.

The microfilariae are transferred into a vector, which are most commonly mosquito species of the genera Culex, Anopheles, Mansonia, and Aedes. Inside the mosquito, the microfilariae mature into motile larvae called juveniles; these migrate to the labium after a period around 10 days. When the infected mosquito has its next blood meal, W. bancrofti larvae are deposited from the mouthparts onto the skin of the prospective host and migrate through microcuts in the dermis or the tract created by the proboscis into the blood stream of the new human host. The larvae move through the lymphatic system to regional lymph nodes, predominantly in the legs and genital area. The larvae develop into adult worms over the course of a year, and reach sexual maturity in the afferent lymphatic vessels. After mating, the adult female worm can produce thousands of microfilariae that migrate into the bloodstream. A mosquito vector can bite the infected human host, ingest the microfilariae, and thus repeat the lifecycle. The organism notably does not multiply within its intermediate host, the mosquito.

Diagnosis

A blood smear is a simple and fairly accurate diagnostic tool, provided the blood sample is taken during the period in the day when the juveniles are in the peripheral circulation. Technicians analyzing the blood smear must be able to distinguish between W. bancrofti and other parasites potentially present.

A polymerase chain reaction test can also be performed to detect a minute fraction, as little as 1 pg, of filarial DNA. Some infected people do not have microfilariae in their blood. As a result, tests aimed to detect antigens from adult worms can be used. Ultrasonography can also be used to detect the movements and noises caused by the movement of adult worms. Dead, calcified worms can be detected by X-ray examinations.

Prevention

Prevention focuses on protecting against mosquito bites in endemic regions. Insect repellents and mosquito nets are useful to protect against mosquito bites. Public education efforts must also be made within the endemic areas of the world to successfully lower the prevalence of W. bancrofti infections.

Eradication

The WHO is coordinating an effort to eradicate filariasis. The mainstay of this programme is the mass use of antifilarial drugs on a regular basis for at least five years.

In April 2011, Sri Lanka was certified by the WHO as having eradicated this disease.

Treatment

The severe symptoms caused by the parasite can be avoided by cleansing the skin, surgery, or the use of anthelmintic drugs, such as diethylcarbamazine, ivermectin, or albendazole. The drug of choice is diethylcarbamazine, which can eliminate the microfilariae from the blood and also kill the adult worms with a dose of 6 mg/kg/day for 12 days, semiannually or annually. A polytherapy treatment that includes ivermectin with diethylcarbamazine or albendazole is more effective than either drug alone. Protection is similar to that of other mosquito-spread illnesses; one can use barriers both physical (a mosquito net), chemical (insect repellent), or mass chemotherapy as a method to control the spread of the disease.

Mass chemotherapy should cover the entire endemic area at the same time. This will significantly decrease the overall microfilarial titer in blood in mass, hence decreasing the transmission through mosquitoes during their subsequent bites.

Antibiotic active against the Wolbachia symbionts of the worm have been tested as treatment. Wolbachia-free worms immediately become sterile, and later die prematurely.

Epidemiology

W. bancrofti is responsible for 90% of lymphatic filariasis. Recently, 120 million worldwide cases of lymphatic filariasis were estimated. W. bancrofti largely affects areas across the broad equatorial belt (Africa, the Nile Delta, Turkey, India, the East Indies, Southeast Asia, Philippines, Oceanic Islands, and parts of South America).

The mosquito vectors of W. bancrofti have a preference for human blood; humans are apparently the only animals naturally infected with W. bancrofti. No reservoir host is known, and the disease could, therefore, potentially be eradicated.

Brugia Malayi

Brugia malayi is a nematode (roundworm), one of the three causative agents of lymphatic filariasis in humans. Lymphatic filariasis, also known as elephantiasis, is a condition characterized by swelling of the lower limbs. The two other filarial causes of lymphatic filariasis are Wuchereria bancrofti and Brugia timori, which both differ from B. malayi morphologically, symptomatically, and in geographical extent.

B. malayi is transmitted by Mansonia mosquitoes and is restricted to South and Southeast Asia. It is one of the tropical diseases targeted for elimination by the year 2020 by the World Health Organization, which has spurred vaccine and drug development, as well as new methods of vector control.

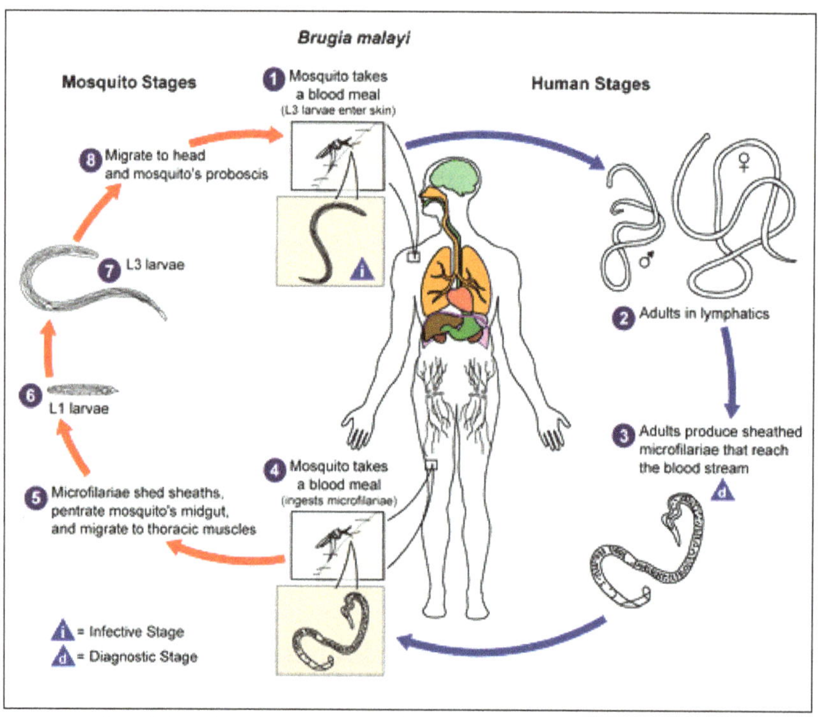

The life cycle of Brugia malayi. Specialty Infectious disease.

Signs and Symptoms

B. malayi is one of the causative agents of lymphatic filariasis, a condition marked by infection and swelling of the lymphatic system. The disease is primarily caused by the presence of worms in the lymphatic vessels and the resulting inflammatory response of the host. Signs of infection are typically consistent with those seen in Bancroftian filariasis—fever, lymphadenitis, lymphangitis, lymphedema, and secondary bacterial infection—with a few exceptions.

Lymphadenitis

Lymphadenitis, the swelling of the lymph nodes, is a commonly recognized symptom of many diseases. It is an early manifestation of filariasis, usually occurs in the inguinal area during B. malayi infection and can occur before the worms mature.

Lymphangitis

Lymphangitis is the inflammation of the lymphatic vessels in response to infection. It occurs early in the course of infection in response to worm development, molting, death, or bacterial and fungal infection. The affected lymphatic vessels become distended and tender, and the overlying skin becomes erythematous and hot. Abscess formation and ulceration of the affected lymph node occasionally occurs during B. malayi infection, more often than in Bancroftian filariasis. Remnants of adult worms can sometimes be found in the ulcer drainage.

Lymphedema

The most obvious sign of infection, elephantiasis, is the enlargement of the limbs–usually the legs. A late complication of infection, elephantiasis is a form of lymphedema and is caused by repeated inflammation of the lymphatic vessels. Repeated inflammatory reactions causes vessel dilation and thickening of the affected lymphatic vessels, which can compromise function. The lymphatic system normally functions to maintain fluid balance between tissues and the blood and serves as an integral part of the immune system. Blockage of these vessels due to inflammatory induced fibrosis, dead worms, or granulomatous reactions can interfere with normal fluid balance, thus leading to swelling in the extremities. Elephantiasis resulting from B. malayi infection typically affects the distal portions of the extremities. Unlike Bancroftian filariasis, B. malayi rarely affects genitalia and does not cause funiculitis, orchitis, epididymitis, hydrocele, or chyluria, conditions more often observed with Bancroftian infection.

Secondary Bacterial Infection

Secondary bacterial infection is common among patients with filariasis. Compromised immune function due to lymphatic damage in addition to lymph node ulcerations and abscesses exposure and impaired circulation due to elephantiasis can cause secondary bacterial or fungal infection. Elephantiasis, in addition to the physical burden of a swollen limb, can be a severely dehabilitating condition given bacterial infection. Part of the WHO's "Strategy to Eliminate Lymphatic Filariasis" targets hygiene promotion programs in order to alleviate the suffering of affected individuals.

However, clinical manifestations of infection are variable and depend on several factors, including host immune system, infectious dose, and parasite strain differences. Most infections appear asymptomatic, yet vary from individual to individual. Individuals living in endemic areas with microfilaremia may never present with overt symptoms, whereas in other cases, only a few worms can exacerbate a severe inflammatory response.

The development of the disease in humans, however, is not well understood. Adults typically develop worse symptoms, given the long exposure time required for infection. Infection may occur during childhood, but the disease appears to take many years to manifest. The incubation period for infection ranges from 1 month to 2 years and typically microfilariae appear before overt symptoms. Lymphedema can develop within six months and development of elephantiasis has been reported within a year of infection among refugees, who are more immunologically naive. Men tend to develop worse symptoms than women.

Causes

Life Cycle

Development and replication of B. malayi occurs in two discrete phases: in the mosquito vector and in the human. Both stages are essential to the life cycle of the parasite.

- Mosquito: The mosquito serves as a biological vector and intermediate host – it is required for the developmental cycle and transmission of B. malayi. The mosquito takes a human blood meal and ingests microfilariae (worm-like sheathed eggs) that circulate in the human blood stream.

In the mosquito, the microfilariae shed sheaths, penetrate the midgut, and migrate to the thoracic muscles were the microfilariae increase in size, molt, and develop into infective larvae (L1 and L3) over a span of 7–21 days. No multiplication or sexual reproduction of microfilariae occurs in the mosquito. The infective larvae (L3) migrate to the salivary glands, enter the proboscis and escape onto human skin when the mosquito takes another blood meal.

- Human: B. malayi undergoes further development in the human as well as sexual reproduction and egg production. The infective larvae (L3) actively penetrate the skin through the bite hole and develop into adults in the lymphatic system over a span of 6 months. Adult worms can survive in the lymphatic system for 5–15 years.

The male and female adult worms mate and the females produce an average of 10,000 sheathed eggs (microfilaria) daily The microfilariae enter the blood stream and exhibit nocturnal periodicity and subperiodicity. Another mosquito takes a blood meal and ingests the microfilariae. Infection depends on the mosquito taking a blood meal during a periodic episode – when microfilariae are present in the bloodstream.

Morphology

Adult

Adult worms resemble typical nematode roundworms. Long and threadlike, B. malayi and other nematode possess only longitudinal muscles and move in an S-shape motion. Adults are typically smaller than adult W. bancrofti, though few adults have been isolated. Female adult worms (50 mm) are larger than male worms (25 mm).

Microfilariae

B. malayi microfilariae are 200–275 μm in length and have a round anterior end and a pointed posterior end. The microfilariae are sheathed, which stains heavily with Giemsa. The sheath is actually the egg shell, a thin layer that surrounds the egg shell as the microfilariae circulates in the bloodstream. The microfilariae retain the sheath until they are digested in the mosquito midgut.

B. malayi microfilariae resemble W. bancrofti and Loa loa microfilariae with minor differences that can aid in laboratory diagnosis. B. malayi microfilariae can be distinguished by the noncontinuous row of nuclei found in the tip of the tail. There are two terminal nuclei that are distinctly separated from the other nuclei in the tail, whereas the tail of W. bancrofti contains no nuclei and Loa loa microfilariae nuclei form a continuous row in the tail. B. malayi microfilariae also have a characteristic cephalic space ratio of 2:1.

Spread

B. malayi is transmitted by a mosquito vector. The principal mosquito vectors include Mansonia, Anopheles, and Aedes mosquitoes. The geographical distribution of the disease is thus dependent on suitable mosquito breeding habitat.

- The nocturnal periodic form is transmitted by Mansonia and some anopheline mosquitoes in open swamps and rice growing areas. These mosquitoes tend to bite at night and appear to only infect humans. Natural animal infections are rare and experimental animals do not retain infection.
- The nocturnal subperiodic form is transmitted by Mansonia in forest swamps, where mosquitoes bite in the shade at any time. Natural zoonotic infections are common. Cats, dogs, monkeys, slow lorises, civet cats, and hamsters have all been successfully experimentally infected with B. malayi from man and may serve as important reservoirs.

The accumulation of many infective mosquito bites – several hundreds to thousands – is required to establish infection. This is because a competent mosquito usually transmits only a few infective L3 larvae, and less than 10% of those larvae progress through all the necessary molting steps and develop into adult worms that can mate. Thus those at greatest risk for infection are individuals living in endemic areas—short term tourists are unlikely to develop lymphatic filariasis.

Diagnosis

Tender or enlarged inguinal lymph nodes or swelling in the extremities can alert physicians or public health officials to infection.

With appropriate laboratory equipment, microscopic examination of differential morphological features of microfilariae in stained blood films can aid diagnosis—in particular the examination of the tail portion, the presence of a sheath, and the size of the cephalic space. Giemsa staining will uniquely stain B. malayi sheath pink. However, finding the microfilariae on blood films can be difficult because of the nocturnal periodicity of some forms of B. malayi.

PCR based assays are highly sensitive and can be used to monitor infections both in humans and the mosquito vectors. However, PCR assays are time-consuming, labor-intensive and require laboratory equipment. Lymphatic filariasis mainly affects the poor, who live in areas without such resources.

The ICT antigen card test is widely used in the diagnosis of W. bancrofti, but commercial antigens of B. malayi have not been widely available. However, new research developments have identified a recombinant antigen (BmR1) that is both specific and sensitive in the detection of IgG4 antibodies against B. malayi and B. timori in an enzyme-linked immunosorbent assay and an immunochromatographic rapid dipstick (Brugia Rapid) test. However, it appears that immunoreactivity to this antigen is variable in individuals infected with other filarial nematodes. This research has led to the development of two new rapid immunochromatographic IgG4 cassette tests – WB rapid and panLF rapid – which detect Bancroftian filariasis and all three species of lymphatic filariasis, respectively, with high sensitivity and selectivity.

Prevention

Vaccines

There is currently no licensed vaccine to prevent lymphatic filariasis. However, recent research has produced vaccine candidates with good results in experimental animals. A glutathione-S-transferase, a detoxification enzyme in parasites isolated from Setaria cervi, a bovine filarial parasite, reduced B. malayi adult parasite burden by more than 82% for 90 days after treatment.

Vector Control

Vector control has been effective in virtually eliminating lymphatic filariasis in some regions, but vector control combined with chemotherapy produces the best results. It is suggested that 11 to 12 years of effective vector control may eliminate lymphatic filariasis. Successful methods of B. malayi vector control include residual house spraying using DDT and insecticide treated bednets. Mansonia larvae attach their breathing tubes to underwater roots and plants in order to survive. While chemical larvicides have only provided partial control, plant removal could prevent vector development, but also potential adverse effects on the aquatic environment. Lymphatic filariasis vector control is neglected in comparison to the far more established efforts to control malaria and dengue vectors. Integrated vector control methods should be applied in areas where the same mosquito species is responsible for transmitting multiple pathogens.

Management

The Global Alliance to Eliminate Lymphatic Filariasis was launched by the World Health Organization in 2000 with two primary goals: 1) to interrupt transmission and 2) to alleviate the suffering of affected individuals. Mass drug treatment programs are the main strategy for interrupting parasite transmission, and morbidity management, focusing on hygiene, improves the quality of life of infected individuals.

Drugs

A goal of community-based efforts is to eliminate microfilariae from the blood of infected individuals in order to prevent transmission to the mosquito. This is primarily accomplished through the use of drugs. The treatment for B. malayi infection is the same as for Bancroftian filariasis. Diethylcarbamazine has been used in mass treatment programs as an effective microfilaricidal drug in several locations, including India. While diethylcarbamazine tends to cause adverse reactions like immediate fever and weakness, it is not known to cause any long-term adverse drug effects. It has been shown to kill both adult worms and microfilariae. In Malaysia, diethylcarbamazine dosages (6 mg/kg weekly for 6 weeks; 6 mg/kg daily for 9 days) reduced microfilariae by 80% for 18–24 months after treatment in the absence of mosquito control. Microfilariae numbers slowly return many months after treatment, thus requiring multiple drug doses over time in order to achieve long-term control. However, it is not known how many years of mass drug administration is required to eliminate transmission. there have been any confirmed cases of diethylcarbamazine resistance as of 2007.

Single doses of two drugs (albendazole-diethylcarbamazine and albendazole-ivermectin) have been shown to remove 99% of microfilariae for a year after treatment and help to improve elephantiasis during early stages of the disease. Ivermectin does not appear to kill adult worms but serves as a less toxic microfilaricide.

Since the discovery of the importance of Wolbachia bacteria in the life cycle of B. malayi and other nematodes, novel drug efforts have targeted the endobacterium. Tetracyclines, rifampicin, and chloramphenicol have been effective in vitro by interfering with larvae molting and microfilariae development. Tetracyclines have been shown to cause reproductive and embryogenesis abnormalities in the adult worms, resulting in worm sterility. Clinical trials have demonstrated the successful reduction of Wolbachia and microfilariae in onchocerciasis and W. bancrofti infected patients. These antibiotics, while acting through a slightly more indirect route, are promising antifilarial drugs.

Hygiene

Secondary bacterial infection is often observed with lymphatic filariasis. Rigorous hygiene practices, including washing with soap and water daily and disinfecting wounds can help heal infected surfaces, and slow and potentially reverse existing tissue damage. Promoting hygiene is essential for lymphatic filariasis patients given the compromised immune and damaged lymphatic systems and can help prevent suffering and disability.

Epidemiology

B. malayi infects 13 million people in south and southeast Asia and is responsible for nearly 10% of the world's total cases of lymphatic filariasis. B. malayi infection is endemic or potentially endemic in 16 countries, where it is most common in southern China and India, but it also occurs in Indonesia, Thailand, Vietnam, Malaysia, the Philippines, and South Korea. The distribution of B. malayi overlaps with W. bancrofti in these regions, but does not coexist with B. timori. Regional foci of endemicity are determined in part by the mosquito vectors.

In 2007, scientists sequenced the genome of Brugia malayi. Identifying the genes of this organism might lead to development of new drugs and vaccines.

To decipher the genome, "Whole Genome Shotgun Sequencing" was performed. The genome was found to be approximately 90-95 megabases in size. The results of the sequencing were then compared to those of the reference nematode Caenorhabditis elegans, along with its prototype Caenorhabditis briggsae. These two free-living nematodes were incorporated in the study and were important for several reasons:

- Comparing genomes using C. elegans allowed for the identification of similar linkages in genes.
- The researchers found a genomic conservation with an absence of conservation at a more local gene level.
- This demonstrated that rearrangements had occurred over time between the C. elegans and B. malayi and allowed researchers to identify genes or proteins that were specific to B. malayi.
- These unique genes were significant because they could have led to the parasitism seen in B. malayi, and therefore are potential targets for future studies.
- Gene Linkages offer insight into the evolutionary trend of parasitic genes that could yield clues to further explain their unique ability to successfully survive for many years in humans.

New Medications

Sequence comparisons between the two genomes allow for mapping of C. elegans orthologs to B. malayi genes. By using orthology mappings (between C. elegans and B.malayi) and incorporating the extensive genomic and functional genomic data, including genome-wide RNAi screens that already exist for C. elegans, potentially essential genes in B. malayi can be identified. Scientists are hoping to be able to target these genes as potential new targets for drug treatments. The longevity of this parasite complicates treatment because most existing drugs target the larvae and thus do not kill adult worms. The drugs often must be taken periodically for years, and the worms can cause a massive immune reaction when they die and releases foreign molecules in the body. Drug treatments for filariasis have not changed significantly in over 20 years, and with the risk of resistance rising, there is an urgent need for the development of new anti-filarial drug therapies. From the genome sequence, several metabolic pathways have been identified, containing dozens of gene products that may be helpful in the discovery of more targeted and effective drug therapies.

Possible new drug targets include:

- Molting,
- Nuclear receptors,
- Collagens and collagen processing,
- Neuronal signaling,
- The B. malayi kinome,
- Reliance on host (B. malayi) and endosymbiont (Wolbachia) metabolism.

These potential new targets for drugs or vaccines may provide new opportunities for understanding, treating and preventing elephantiasis.

Relationship with Wolbachia

The relationship between the Wolbachia bacteria and B. malayi is not fully understood. Extrapolating from research done with Wuchereria bancrofti, another nematode that causes filariasis, Wolbachia may aid in embryogenesis of the worm, be responsible for potent inflammatory responses from macrophages and filarial disease, and be linked to the onset of lyphedema and blindness sometimes associated with B. malayi infections. In a study done by the University of Bonn in Ghana, doxycycline effectively depleted Wolbachia from W. bancrofti. It is likely that the mechanism of doxycycline is similar to that in other filarial species, i.e., a predominant blockade of embryogenesis, leading to a decline of microfilariae according to their half-life. This could render doxycycline treatment an additional tool for the treatment of microfilaria-associated diseases in Bancroftian filariasis, along with B. malayi fiariasis. The course of treatment with doxycycline could be much shorter as it would make the adult worm sterile in one shot rather than repeatedly have to target the replenished larvae that current treatments kill, and there would be fewer side effects for the infected individual.

Transplant Research

Another hopeful use for the research is in the area of transplant research. Because the B. malayi genome is the first parasitic genome to have been sequenced, the implications on the mechanism of parasitism in humans are crucial to understand. According to Alan L. Scott, Ph.D., a collaborator at Johns Hopkins University, understanding how a particular parasite, such as B. malayi, can adapt to humans, may yield medical benefits far beyond treating elephantiasis. According to the author, "This worm can reside in the host for years and not necessarily cause disease, in fact the less disease the individual has, the more worms there are in circulation. Now that we know those genes don't exist in humans we can target them to control disease." Some of the predicted proteins for these new genes appear similar to known immuno-modulator proteins, regulators of the immune system, suggesting that they are involved in deactivating the host's immune system to ensure the parasite remains undetected. Knowledge of these previously unknown immune suppressors could also be of use in organ transplants and to help treat autoimmune disease.

According to the Filarial Genome Project being done by The Special Programme for Research and Training in Tropical Diseases, the Brugia malayi MIF gene is expressed in all life-cycle stages of the parasite, and results suggest that B. malayi MIF may interact with the human immune system during the course of infection by altering the function of macrophages in the infected individual. Studies are currently testing the hypothesis that MIF may be involved in reducing the host's immune response to the microfilariae. Understanding how this particular parasite has adapted to humans may help organ transplant researchers by figuring out how to prevent the immune system from attacking the transplanted tissue.

Loa Loa

Loa loa is the filarial nematode (roundworm) species that causes Loa loa filariasis. Loa loa actually means "worm worm", but is commonly known as the "eye worm", as it localizes to the conjunctiva of the eye. Loa loa is commonly found in Africa. It mainly inhabits rain forests in West Africa and has native origins in Ethiopia. The disease caused by Loa loa is called loiasis and belongs to the so-called neglected diseases.

L. loa is one of three parasitic filarial nematodes that cause subcutaneous filariasis in humans. The other two are Mansonella streptocerca and Onchocerca volvulus (causes river blindness).

Maturing larvae and adults of the "eye worm" occupy the subcutaneous layer of the skin – the fat layer – of humans, causing disease. The L. loa adult worm which travels under the skin can survive up to 10–15 years, causing inflammations known as Calabar swellings. The adult worm travels under the skin, where the female deposits the microfilariae which can develop in the host's blood within 5 to 6 months and can survive up to 17 years. The young larvae, or microfilariae, develop in horseflies of the genus Chrysops (deer flies, yellow flies), including the species C. dimidiata and C. silacea, which infect humans by biting them. After bites from these infected flies, the microfilariae are unique in that they travel in the peripheral blood during the day and migrate into the lungs at night.

Morphology

L. loa worms have a simple body consisting of a head that lacks lips, a body, and a blunt tail. The outer body of the worm is composed of a cuticle with 3 main layers made up of collagen and other compounds which aid in protecting the nematodes while they are inside the digestive system of their host. Juveniles have a similar appearance to adult worms, but are significantly smaller. Male adults range from 20 to 34 mm long and 350 to 430 µm wide. Female adults range from 20 to 70 mm long and can be about 425 µm wide. They vary in color.

Lifecycle

The human is the definitive host, in which the parasitic worms attain sexual maturity, mate, and produce microfilariae. The flies serve as intermediate hosts in which the microfilariae undergo part of their morphological development, and then are borne to the next definitive host.

Two species of Chrysops deerflies, C. silacea and C. dimidiata, are the main vectors for this filariasis.

- A fly bearing third-stage filarial larvae in its proboscis infects the human host through the bite wound.
- After entering the human host, the larvae mature into adults, commonly in subcutaneous tissue. Adult females measure about 40 to 70 mm in length and 0.5 mm in diameter. Males measure some 30 to 34 mm by 0.35 to 0.43 mm.
- The adult female produces large numbers of microfilariae, about 250 to 300 µm in length and 6-8 µm in width. She continues to do so continuously for her lifetime, which typically spans several years.
- Microfilariae tend to reside within spinal fluids, urine, and sputum; by day, they also circulate in the bloodstream. Apart from their presence in bodily fluids, however, microfilariae in the noncirculation phase also occur in the lungs.

- The vector fly ingests microfilariae while feeding on the host's blood.
- Once inside the vector, the microfilaria sheds its sheaths and escapes through the walls of the midgut into the fly's haemocoel.
- It then migrates through the haemolymph into the wing muscles in the fly's thorax.
- In the thoracic muscles, the microfilaria develops successively into a first-stage larva, second-stage larva, and finally into the infectious third-stage larva.
- The third-stage larva migrates to the fly's proboscis.
- Once the larva is established in the proboscis and the fly takes its next human blood meal, the cycle of infection continues.

Disease

Signs and Symptoms of Loa Loa

Usually, about 5 months are needed for larvae (transferred from a fly) to mature into adult worms, which they can only do inside the human body. The most common display of infection is the localized allergic inflammations called Calabar or Cameroon swellings that signify the migration of the adult worm in the tissues away from the injection site by the vector. The migration does not cause significant damage to the host and is referred to as benign. However, these swellings can be painful, as they are mostly found near the joints.

Although most infections with L. loa are asymptomatic, symptoms generally do not appear until years, or even more than a decade, after the bite from an infected fly. However, symptoms can appear as early as 4 months after a bite. These parasites have a diurnal periodicity in which they circulate in the peripheral blood during the daytime, but migrate to vascular parts of the lungs during the night, where they are considered noncirculatory. Therefore, the appearing and disappearing characteristics of this parasite can cause recurrent swelling that can cause painful enlargements of cysts in the connective tissue surrounding tendons. Additionally, chronic abscesses can be caused by the dying worms.

The most visual sign of an adult worm infections is when the worm crosses the sclera of the eye, which causes significant pain to the host and is usually associated with inflammation and less likely, blindness. Eye worms typically cause little eye damage and last a few hours to a week. Other tissues in which this worm can be found includes: the penis, testes, nipples, bridge of the nose, kidneys, and heart. The worms in these locations are not always possible to visualize.

Risk Factors of Loa Loa

People at the highest risk for acquiring loiasis are those who live in the rainforests of West or Central Africa. Furthermore, the probability of getting bitten by a deer or mango fly increases during the day and during rainy seasons. The flies are also attracted to smoke from wood fires. These flies do not commonly enter houses, but they are attracted to the houses that are well lit, so will congregate outside.

Travelers can be infected in less than 30 days after arriving in an affected area, although they are more likely to be infected whilst being bitten by multiple deerflies over the course of many months. Men are more susceptible than women due to their increased exposure to the vectors during activities such as farming, hunting, and fishing.

Diagnosis

The main methods of diagnosis include the presence of microfilariae in the blood, the presence of a worm in the eye, and the presence of skin swellings. However, in cases where none of those is the case, a blood count can be done. Patients with infections have a higher number of blood cells, namely eosinophils, as well as high IgE levels that indicate an active infection with helminth parasites. Due to the migration of microfilariae during the day, the accuracy of a blood test can be increased when samples are taken between 10 am and 2 pm. A Giemsa stain is the most commonly used diagnostic test that uses a thick blood smear to count the microfilariae. Other than blood, microfilariae can also be observed in urine and saliva samples.

Treatment

Adult worms found in the eye can be surgically removed with forceps after being paralyzed with a topical anesthesia. The worm is not paralyzed completely, so if it is not extracted quickly, it can vanish upon attempting extraction.

Ivermectin has become the most common antiparasitic agent used worldwide, but can lead to residual microfilarial load when given in the management of loiasis. Treatment with ivermectin has shown to produce severe adverse neurological consequences in some cases. These treatment complications can be increased in individuals co-infected with onchocerciasis. Some of these patients experienced cases of coma and resultant encephalopathy, parkinsonism, and death. After about 12 hours, the first signs start to appear and include fatigue, pain in joints, mutism, and incontinence. Severe disorders of the consciousness start to develop after about a day.

Although Ivermectin is a common treatment for loiasis, the Center for Disease Control (CDC) recommends treatment with diethylcarbamazine (DEC). Symptoms may be resolved with as little as 1-2 courses of DEC. DEC is chosen over Ivermectin because evidence supports its ability to kill both the adult worms and the microfilariae, which are the main cause of the severe neurological problems mentioned above. In some cases, albendazole may also be an effective treatment used to reduce the microfilariae prior to treatment with DEC. The body's response to albendazole is slow, so the patient being treated must be monitored closely and frequently to ensure it is effective.

Prevention of Loa Loa

Currently, no control programs or vaccines for loiasis are available. However, diethylcarbamazine treatment is suggested to reduce risk of infection. Avoiding areas where the vectors, deer and mango flies, are found also reduces risk. This includes swamps, bogs, and shaded areas near rivers or near wood fires. Fly bites can be reduced by using insect repellents such as DEET and wearing long sleeves and pants during the daytime. Permethrin treatment on clothes is an additional repellent that could be used. Also, using malaria nets can reduce the number of fly bites acquired.

Epidemiology

Reports of microfilaremia have been made in Angola, Benin, Cameroon, Central African Republic, Congo, the Democratic Republic of Congo, Equatorial Guinea, Gabon, Nigeria, and Sudan, and possibly rare cases in Chad, Ghana, Guinea, Liberia, Uganda, and Zambia. Of the 10 countries that have high rates of infection, about 40% of the people who live in the area have reported being infected with the worm in the past. The population in high-risk areas is about 14.4 million; in addition, 15.2 million people live in areas where around 20–40% of people admitted to having the worm in the past.

Epidemiological studies have been emphasized in the western part of Africa. In this area, the disease is considered endemic. A study conducted by the Research Foundation in Tropical Diseases and Environment in 2002 had a sample of 1458 persons, spanning 16 different villages, and found Loa loa presence in these villages ranging from 2.22 to 19.23% of the population. The disease was found to be slightly more prevalent in men.

In a different country in western Africa, a cross-sectional community survey was conducted in Gabon, Africa. The study was performed by the department of Tsamba-Magotsi from August 2008 to February 2009. The study of 1,180 subjects evaluated the presence of microfilaria using microscopy. The carriage rate of L. loa in the subjects tested was 5%. This rate falls within the range of the study listed above.

In the western part of Africa, an increase in prevalence has been associated with the distribution of ivermectin, which is used to prevent the infection of onchocerciasis, which is also very prevalent in the same region. Patients with L. loa who are treated with ivermectin have extreme adverse effects, including death. Therefore, a prevalence mapping system was created called REMO. REMO is used to determine which areas to distribute the ivermectin based on lower levels of L. loa prevalence. The area discovered to be the most overlapping was where Cameroon and the Democratic Republic of Congo overlap.

A study performed to review reported cases of L. loa in nonendemic countries in the past 25 years reviewed 102 cases of imported loiasis, 61 of them from Europe and 31 from the USA. Three-quarters of the infestations were acquired in three countries considered endemic: Cameroon, Nigeria, and Gabon. In the subjects viewed, peripheral blood microfilariae were detected 61.4% of the time, eye worm migration 53.5% of the time, and Calabar swellings 41.6% of the time. A trend appeared in the symptoms of the patients where Calabar swellings and eosinophilia were more common among travelers. African immigrants tended to have microfilaremia. Eye worm migration was observed in a similar proportion between the two groups. Only 35 of the patients underwent clinical follow-up. The researchers concluded that L. loa would end up migrating to Europe and the United States, due to increased travel to already endemic regions.

Mansonella Ozzardi

Mansonella ozzardi is strictly a New World parasite that has never been found in the Old World. It was first studied in blood samples taken from Carib indians in British Guyana. The organism is located throughout northern Argentina, the Amazon drainage, the northern coast of South America, Central America and several islands of the West Indies.

Habitat

Mansonella ozzardi is concentrated in the tropical regions of the New World. Like all parasitic species, M. ozzardi requires nutrients from its hosts, and as an endoparasite it also requires the host to provide the habitat for it to survive. Adults live in the human body cavity among the mesenteries, peritoneum, and in the subcutaneous tissue, while juvenile stages require development in a Dipteran host. The immature worms migrate to the circulatory system of the human host, waiting to be ingested with an insect's blood meal. The factors that determine the suitability of particular Dipteran hosts has not yet been discovered, but after development is complete, the worms re-enter the human host when an infected insect bites another human. Thus, the entire life cycle of this organism is spent inside another species.

M. ozzardi may be especially well adapted to aboriginal Indians, in which the prevalence of infection is extremely high. In 1970, Marenhelle and German reported that 96.2% of adult Indians were infected in a region of Southern Colombia.

Physical Description

Mansonella ozzardi, being a nematode, is cylindrical, bilaterally symmetrical and possesses a pseudocoel, which is a fluid filled body cavity between the digestive tract and the body wall. The cuticle has three main outer non-cellular layers made of collagen and other compounds. The cuticle layer protects the nematodes so they can invade the digestive tracts of animals.

Nematodes have longitudinal muscles along the body wall. The muscles are obliquely arranged in bands. Dorsal, ventral and longitudinal nerve cords are connected to the main body of the muscle.

Adult Mansonella ozzardi are long and slender with reduced lips. Females are larger than males, and produce thousands of offspring called microfilariae, which are between 185-200 um long. Microfilariae of this species are unsheathed. When stained, the presence or absence of a sheath, internal nuclei and organs can all be seen, with the organization of these aiding in identification and classification of the different filarial worm species.

Development of Mansonella Ozzardi

Once in humans, the adults mate and females produce unsheathed microfilariae, which have a sharp tail. Females of this species are ovoviviparous, which is when embryos develop within the female's body, and the microfilariae produced are not as well differentiated as normal first stage juveniles. At this stage they are considered advanced embryos. The worms molt four times, the first two before hatching, and then before their adult stage.

Reproduction

Females may produce a phermomone to attract males. The male coils around a female with his curved area over the female genital pore. The gubernaculum, made of cuticle tissue, guides spicules which extend through the cloaca and anus. Males use spicules to hold the female during copulation. Nematode sperm are amoeboid-like and lack flagella.

Behavior

The microfilariae wandering through the blood are ingested via a blood meal by species of biting midges of the genus Culicodes and black flies of the genus Simulium. The microfilariae are non-periodic, thus they can be found in the peripheral circulatory system at all times. Upon entrance to the insect host, the microfilariae penetrate the body cavity and migrate to the thoracic muscles where development to infective stages takes place. In one study, the time frame in which the microfilariae in Culicoides furens migrated to the thorax was 24 hours. Once located in the thoracic muscles, M. ozzardi lays parallel to the muscle cells and over 2 to 3 days becomes shorter and fatter. After two molts, the larvae become active and migrate out from the thorax and go toward the head. Once here, they move down to the interior of the labium. When stimulated by warmth, the larvae escape the arthropod host when it takes a blood meal from a human. Back inside the human circulatory system, M. ozzardi makes its way to the internal body cavities and continues its growth and development to sexual maturity.

Communication and Perception

Nematodes within the Secernentea have phasmids, which are unicellular glands. Phasmids likely function as chemoreceptors. Females may produce pheromones to attract males.

Nematodes in general have papillae, setae and amphids as the main sense organs. Setae detect motion (mechanoreceptors), while amphids detect chemicals (chemoreceptors).

Food Habits

Mansonella ozzardi is an endoparasitic organism that utilizes several Diptera species as an intermediate hosts and humans as a definitive host (the host in which sexually maturity is reached). Both entrance and exit from the definitive host occurs through the skin. Adults live in the human body cavity among the mesenteries, peritoneum, and in the subcutaneous layers feeding on the host's tissue fluid to grow. Infective larvae develop in the thoracic muscles of the arthropod host, once again feeding on the surrounding tissues to grow and molt.

Pharyngeal glands and intestinal epithelium produce digestive enzymes to feed on the hosts' body fluids. Extracellular digestion begins within the lumen and is finished intracellularly.

Predation of Mansonella Ozzardi

These parasites are probably not preyed on directly, but are ingested from host to host. Larval mortality is high as most of the parasites do not reach appropriate hosts.

Ecosystem Roles

Mansonella ozzardi is an endoparasitic organism that utilizes several species of Diptera as an intermediate host and humans as a definitive host.

Economic Importance for Humans

The effects of Mansonella worms in a human can be profound but most infected hosts are asymptomatic. Increased age has been correlated with increased worm levels in the blood, which indicates that the most likely individuals to experience symptoms associated with M. ozzardi infection are those over the age of 25. Although most individuals are symptom-less, joint pains, headaches, coldness of the legs, and itchy red spots have been described in conjunction with M. ozzardi infection. Mansonella ozzardi is one of several filarial worm species associated with lymphatic filariasis. In addition, the manifestations of the parasite can mimic the more serious bancroftian filariasis, with polylymphadenitis, lymphedema, elephantiasis, and hepatomegaly over time.

The treatment of filarid infestations can vary. Conventional treatment has been with diethylcarbamazine (DEC) but recent studies have shown that Ivermectin can be an effective treatment of M. ozzardi depending on initial microfilariae levels. Also, prevention of the worms entering a human host relies heavily on insecticides and insect repellents. This action helps prevent infected Dipteran hosts from biting humans.

Giardia Lamblia

Giardia lamblia (syn. Giardia intestinalis, Giardia duodenalis) is a flagellated unicellular eukaryotic microorganism that commonly causes diarrheal disease throughout the world. It is the most common cause of waterborne outbreaks of diarrhea in the United States and is occasionally seen as a cause of food-borne diarrhea. In developing countries, there is a very high prevalence and incidence of infection, and data suggest that long-term growth retardation can result from chronic giardiasis. In certain areas of the world, water contaminated with G. lamblia cysts commonly causes travel-related giardiasis in tourists.

Giardia species have two major stages in the life cycle. Infection of a host is initiated when the cyst is ingested with contaminated water or, less commonly, food or through direct fecal-oral contact. The cyst is relatively inert, allowing prolonged survival in a variety of environmental conditions. After exposure to the acidic environment of the stomach, cysts excyst into trophozoites in the proximal small intestine. The trophozoite is the vegetative form and replicates in the small intestine, where it causes symptoms of diarrhea and malabsorption. After exposure to biliary fluid, some of the trophozoites form cysts in the jejunum and are passed in the feces, allowing completion of the transmission cycle by infecting a new host.

The clinical aspects of giardiasis have been reviewed recently, as has the host immune response to giardiasis and epidemiology. This review will focus primarily on the biology of the organism and deal relatively little with the clinical disease or the host-parasite interaction. Since a previous review of Giardia species, considerable progress has been made in the understanding of the organism, and these new advances will be emphasized.

Classification and Evolution of Giardia Species

An appropriate classification for Giardia spp. is critical to an understanding of the pathogenesis and epidemiology of infection, as well as the biology of the organism. This process has been difficult for a number of reasons. (i) The (presumed) asexual nature of the organism does not allow mating experiments to allow species designation. For clonal organisms in the same clade, there are no well-defined criteria for species designation; these designations remain controversial. (ii) Many of the earlier descriptions of Giardia spp. assumed a different species for each host and consequently overestimated the number of species. Subsequent species descriptions based on morphologic differences detected by light microscopy have probably underestimated the differences among isolates, strains, or species. (iii) Cross-transmission experiments of Giardia from one host to another have yielded inconsistent results. (iv) The available tools for distinguishing Giardia isolates have been inadequate until the recent introduction of molecular and electron micrographic techniques for classifying Giardia spp. In view of these concerns, a review of the history of the description of Giardia and the designation of Giardia species is warranted.

Giardia was initially described by van Leeuwenhoek in 1681 as he was examining his own diarrheal stools under the microscope. The organism was described in greater detail by Lambl in 1859, who thought the organism belonged to the genus Cercomonas and named it Cercomonas intestinalis. Thereafter, some have named the genus after him while others have named the species of the human form after him (i.e., G. lamblia). In 1879, Grassi named a rodent organism now known to be a Giardia species, Dimorphus muris, apparently unaware of Lambl's earlier description. In 1882 and 1883, Kunstler described an organism in tadpoles that he named Giardia, the first time Giardia was used as a genus name. In 1888, Blanchard suggested the name Lamblia intestinalis, which Stiles then changed to G. duodenalis in 1902. Subsequently, Kofoid and Christiansen proposed the names G. lamblia in 1915 and G. enterica in 1920. There continued to be controversy about the number of Giardia species for many years, with some investigators suggesting species names on the basis of host of origin and others focusing on morphology. For example, over 40 species names had been proposed on the basis of host of origin. Simon, on the other hand, used morphologic criteria to distinguish between G. lamblia and G. muris and accepted the name G. lamblia for the human form. In 1952, Filice published a detailed morphologic description of Giardia and proposed that three species names be used on the basis of the morphology of the median body: G. duodenalis, G. muris, and G. agilis. The species name G. lamblia became widely accepted through the 1970s. Since the 1980s, some have encouraged the use of the name G. duodenalis, and in the 1990s, the name G. intestinalis has been encouraged by other investigators. At this time there does not appear to be adequate reason to abandon the term G. lamblia, which has been widely accepted in the medical and scientific literature.

G. lamblia is pear shaped and has one or two transverse, claw-shaped median bodies; G. agilis is long and slender and has a teardrop-shaped median body; and the G. muris trophozoite is shorter and rounder and has a small, rounded median body. G. lamblia is found in humans and a variety of other mammals, G. muris is found in rodents, and G. agilis is found in amphibians.

Giardia Species

Species name	Hosts	Morphology by:		Molecular data
		Light microscopy	Electron microscopy	
G. agilis	Amphibians	Long and slender; teardrop-shaped median body	Not Available	Not Available

G. muris	Rodents	Short and rounded; small rounded median body	Not Available	Distant from G. lamblia
G. lamblia	Numerous mammals, including humans	Pear shaped; one or two transverse, claw-shaped median bodies	Not Available	Clade with multiple genotypes
G. ardeae	Herons	Same as G. lamblia	Ventral disk and caudal flagellum similar to G. muris	Closer to G. lamblia than to G. muris
G. psittaci	Psittacine birds	Same as G. lamblia	Incomplete ventrolateral flange, no marginal groove	Not Available
G. microti	Voles and muskrats	Same as G. lamblia	Cysts contain two trophozoites with mature ventral disks	Similar to G. lamblia genotypes

For the Giardia isolates grouped with G. lamblia on the basis of morphologic criteria discernible by light microscopy, differences that can be detected by electron microscopy have allowed the description of additional species, G. psittaci from parakeets and G. ardeae from herons. Another species, Giardia microti, has been suggested on the basis of host specificity for voles and muskrats, differences in the cyst as assessed by electron micrography, and by differences of the 18S rRNA sequences compared with G. lamblia of human origin.

Genotypes of G. Lamblia

Molecular classification tools have been of great value in understanding the pathogenesis and host range of Giardia isolates obtained from humans and a variety of other mammals. The first study of the molecular differences of G. lamblia isolates was a zymodeme analysis of five axenized isolates, three from humans, one from a guinea pig, and one from a cat, using six metabolic enzymes. Zymodeme analysis consists of the typing of organisms based on the migration of a set of enzymes on a starch gel in the presence of an electric field. The migration depends on the size, structure, and isoelectric point of these enzymes. Since these properties are a function of the primary amino acid sequence, differences in the zymodemes should reflect differences in the sequences of the genes encoding these enzymes. In 1985, restriction fragment length polymorphism analysis of 15 isolates was performed using random probes. These studies resulted in the description of three groups; group 3 was so different from groups 1 and 2 that the suggestion of a separate species designation was made. Subsequently, a number of other molecular classification studies have been performed using zymodeme analysis and restriction fragment length polymorphism analysis. Pulsed-field gel electrophoresis (PFGE) chromosome patterns have also been studied but are of limited value for classification because of the frequent occurrence of chromosome rearrangements Likewise, classification by surface antigens is limited by antigenic variation of the variant-specific proteins (VSPs). These studies have been very useful, but the conclusions that can be drawn from these types of data are limited by the semiquantitative nature of the data. To allow a more quantitative comparison of Giardia isolates, sequence comparisons of the small-subunit rRNA, triosephosphate isomerase (tim), and glutamate dehydrogenase (GDH) genes have been utilized in a number of subsequent studies.

These studies have all confirmed the division of G. lamblia human isolates into two major genotypes. The first consists of Nash groups 1 and 2, Mayrhofer assemblage A, groups 1 and 2, and the Polish isolates, while Nash group 3, Mayrhofer assemblage B, groups 3 and 4, and the Belgian isolates form the other major genotype. For example, the tim nucleotide sequences of the group 1 and 2 isolates diverged by 1% in the protein coding region and 2% in the flanking regions, while groups 1 and 3 diverged by 19% and the flanking regions were so dissimilar as to preclude their alignment. The small-subunit or 18S rRNA (SS rRNA) sequence shows a 1% divergence between groups 1 and 3, reflecting its more highly conserved sequence. In addition to their marked genetic differences, the two genotypes may have a number of important biologic differences. For example, the GS isolate (group 3) was significantly more pathogenic in infections of human volunteers than was the WB isolate (group 1). Group 3 organisms also appear to grow much more slowly in axenic culture than do genotype 1 organisms.

Proposed designation	Nash group	Mayrhofer assemblage	Origin	Hosts
Genotype A-1	1	A (group 1)	Poland	Human, beaver, cat, lemur, sheep, calf, dog, chinchilla, alpaca, horse, pig, cow
Genotype A-2	2	A (group 2)		Human, beaver
Genotype B	3	B (groups 3 and 4)	Belgium	Human, beaver, guinea pig, dog, monkey
				Dog
				Dog
		C		
		D		
				Cow, sheep, alpaca, goat, pig
		E (or A-livestock)		Cat
				Rat
		F		
		G		

More recently, a number of additional assemblages (genotypes) have been proposed for Giardia isolates from a variety of mammals. These isolates are morphologically identical to human G. lamblia, but sequences of their protein-coding regions differ. These studies have allowed the identification of a dog isolate that is genetically distinct from human G. lamblia. Giardia isolates from dogs have been notoriously difficult to axenize, in comparison to isolates from humans, leading to the proposal that Giardia isolates from dogs were different from cat or human isolates. However, a few dog isolates have been axenized and characterized. To further evaluate the zoonotic potential of dog Giardia, suckling-mouse infections were established from 11 consecutive infected dogs. On the basis of sequence analysis, these were assigned to two assemblages (C and D) that were quite distinct from assemblages A and B. A PCR-based study of nine fecal isolates from dogs found that one of the nine was similar to human isolates while the other eight were different. These results suggest that most dog isolates are genetically distinct from those found in humans and have little or no potential for zoonotic transmission.

Separate assemblages (E through G) have also been proposed for hoofed livestock, cat, and rat isolates. This same sequence-based study demonstrated that G. microti was a member of this compilation of seven assemblages. Further studies of Giardia obtained from cattle have demonstrated that some of the isolates belong to the livestock assemblage (assemblage E) while others belong to assemblage A (genotype 1) and thus may have the potential for human infection. Assemblages C through G have not yet been isolated from humans, suggesting the likelihood that some genotypes of G. lamblia have a broad range of host specificity that includes humans while others appear to be more restricted in their host range and may not pose a risk of zoonotic transmission. Whether these seven assemblages should be considered separate species should await further data and consensus. Priority should go to the first classification suggested by Nash et al. in 1985. However, "group" has been used in different ways by different authors. In addition, it is clear from data published after the initial description that Nash groups 1 and 2 are closely related while group 3 is genetically distinct enough to allow consideration of a separate species or subspecies name as suggested. The assignment of species status to clonal organisms is problematic and should be delayed until there is general agreement among investigators, but agreement on genotype designation should facilitate a better understanding and coordination of the literature. Therefore, I propose that genotype A be accepted as the designation for Nash group 1 (A-1), Nash group 2 (A-2), assemblage A, and the Polish isolates. Genotype B would then designate Nash group 3, assemblage B, and the Belgian isolates. It may also be appropriate to use a similar designation for assemblages C through G, since these also reflect significant genetic differences as well as host differences.

Host-Parasite Coevolution

Whenever members of a grouping of parasitic organisms parasitize different hosts, it is appropriate to address the question of coevolution of the host and parasite. For Giardia spp., the ability to do so has been limited because of the uncertainty about the host specificity of a specific Giardia species or genotype as well as the problem of appropriate classification of isolates. The development of good molecular classification tools and a better understanding of host specificity makes it reasonable to address the possibility of coevolution of different species or genotypes of Giardia with their hosts. The greater difference between G. lamblia and G. ardeae than among the G. lamblia genotypes supports the idea that the divergence between G. lamblia and G. ardeae accompanied the divergence between birds and mammals. However, G. lamblia and G. ardeae are closer to each other than to G. muris, the opposite of the expected finding, since mice diverged from other mammals more recently than from birds. Sequence information is not yet available from G. agilis to allow a similar comparison of the amphibian and mammalian Giardia species. The closer-than-expected relationship between G. ardeae and G. lamblia could perhaps be explained if transmission of Giardia between birds and mammals was followed by divergence of the two lineages with their hosts.

Giardia and other Diplomonads as Early-branching Eukaryotes

Traditionally, all living organisms have been classified as prokaryotes or eukaryotes, and some still argue for retaining the two major divisions. However, the most widely accepted classification now utilizes three major divisions, Archaea (archaebacteria), Bacteria (eubacteria), and Eukarya (eukaryotes), which can then be divided into kingdoms. With either classification system, G. lamblia is clearly a eukaryotic organism and has been considered a member of the protozoa, the more "animal-like" of the unicellular eukaryotes. These protozoan organisms have traditionally been classified by their morphology into flagellates, ciliates, amebae (rhizopods), and sporozoa. Thus, G. lamblia was classified with the flagellated protozoans, including the kinetoplastids (e.g., Leishmania spp. and Trypanosoma spp.), parabasalids (e.g., Trichomonas vaginalis), and Dientamoeba (e.g., Dientamoeba fragilis). Giardia has been placed in the order Diplomonadida (two karyomastigonts, each with four flagella, two nuclei, no mitochondria, and no Golgi complex; cysts are present, and it can be free-living or parasitic) and the family Hexamitidae (six or eight flagella, two nuclei, bilaterally symmetrical, and sometimes axostyles and median or parabasal bodies), along with the mole parasite Sppironucleus muris and the free-living organism Hexamita inflata. Some of the higher orders of classification do not appear to be phylogenetically valid, such as the placement of all flagellated protozoans together. However, the family Hexamitidae does appear to be a monophyletic group.

One recent classification system with one prokaryotic and five eukaryotic kingdoms has retained Protozoa as one of the six kingdoms, but most recent proposals have suggested abandoning the term "protozoa" in favor of the more general but at the same time more precise term "protista". This review will use the term "protista," but it should be noted that in the clinical context, G. lamblia is usually referred to as a protozoan.

Recent classifications of the eukaryotic microbial organisms have depended primarily on molecular comparisons. An ideal molecular classification system would be based on a gene(s) that is required for all life and that is sufficiently highly conserved across all forms of life that accurate alignments allow comparison and classification of all organisms. In many ways, rRNA has been the most useful gene for molecular comparisons, because rRNA sequences are highly conserved across life and because the function of the rRNA is so central to the biology of the organism. Therefore, the most widely accepted classification scheme has been based on SS (18S) rRNA sequences. Based on comparisons of SS rRNA sequences, G. lamblia was proposed as one of the most primitive eukaryotic organisms, along with T. vaginalis and the microsporidia. The use of SS rRNA sequences to place Giardia as an early-branching eukaryote has been criticized because of the high G+C content of the SS rRNA of Giardia (75%) and because Giardia spp. are parasitic organisms; artifacts may be introduced by high rates of mutation accompanying host adaptation. An analysis of the early-branching eukaryotes suggested that the basal position of Giardia was an artifactual result of long-branch attraction due to a greater evolutionary rate of Giardia. Analysis of the large subunit of RNA polymerase II and reanalysis of the eF1 and eF2 sequences also supported the idea of the long-branch artifact. However, no such effect was shown for the eRF3 tree. In addition, the absence in G. lamblia of the highly conserved N-terminal domain of eRF3 found in other eukaryotes including T. vaginalis suggested the divergence of G. lamblia before the acquisition of the N-terminal domain It should also be noted that the phylogenetic placement of Hexamita, a free-living organism in the same family (as determined by morphologic criteria), avoids artifacts due to high G+C content or parasitism. The G+C content of the Hexamita inflata SS rRNA is 51%, and H. inflata was found to be monophyletic with Giardia The classification of H. inflata with G. lamblia is also supported by a comparison of the glyceraldehyde-3-phosphate dehydrogenase sequences of G. lamblia, Trepomonas agilis, H. inflata, and Spironucleus sp. A comparison of the SS rRNA sequences of Giardia, Hexamita, Trepomonas, and Spironucleus, all diplomonads, showed that all were phylogenetically related and that the last three comprised one clade while Giardia occupied another clade. These sequence-based comparisons have led to the following proposed classification system for Giardia: kingdom Protozoa, subkingdom Archezoa (includes the phyla Metamonada and Microsporidia), subphylum Eopharyngia, class Trepomonadea, subclass Diplozoa, and order Giardiida (includes the families Octomitidae and Giardiidae). In this classification system, the diplomonads are considered to comprise the subclass, Diplozoa, and Hexamita, Trepomonas, and Spironucleus are all members of the other diplozoan order, Distomatida.

An examination of the different genes used for the phylogenic classification of Giardia shows that the genes used in transcription and translation generally yield a basal position for Giardia, although artifact due to long-branch attraction has not been ruled out. Interestingly, a phylogenetic tree of one of the genes thought to be of mitochondrial origin (cpn60) also suggests that Giardia is an early-branching eukaryote. Of the other groups of genes, the metabolic genes do not yield a clear position, but some suggest a eubacterial origin for Giardia. Cytoskeletal genes give mixed results, with actin giving an early divergence but tubulin yielding a divergence that is later than that of Entamoeba histolytica. Among other classes of genes, HSP70 and cathepsin B phylogenies suggest an early divergence for Giardia. Thus, although the answers are not conclusive, most of the data suggest that G. lamblia and the other diplomonads are among the most basal of the extant eukaryotes.

Evolution of Eukaryotes

G. lamblia is a typical eukaryotic organism in that it has a distinct nucleus and nuclear membrane, cytoskeleton, and endomembrane system, but it lacks other organelles that are nearly universal in eukaryotes, such as nucleoli and peroxisomes. In addition, G. lamblia is anaerobic, lacking mitochondria or any of the components of oxidative phosphorylation.

The hypothesis that eukaryotes arose through endosymbiotic events, resulting in the origin of plastids and mitochondria and the use of oxidative phosphorylation as the major source of energy production, has become widely accepted. One version of this hypothesis is that a common ancestor of the archaebacteria and eukaryotes evolved a nucleus and cytoskeleton. A descendant then endocytosed a eubacterium, resulting in the development of a mitochondrion. The observation that certain amitochondriate eukaryotes (e.g., G. lamblia and other diplomonads as well as Trichomonas vaginalis) were basal on an 18S rRNA tree led to the suggestion that those organisms descended from Archezoa before the plastid symbiotic event. However, the subsequent recognition of mitochondrial genes in the genomes of T. vaginalis and G. lamblia has led to the suggestion that the amitochondrial protists have secondarily lost their mitochondria.

An alternative version of the symbiotic hypothesis is that eukaryotes developed through a symbiotic association of an archaebacterium with a eubacterium. In this proposal, the eubacterial endosymbiont provided the pathways for organisms depending on the metabolism of pyruvate by mitochondrial respiration as well as those depending on pyruvate metabolism by pyruvate:ferredoxin oxidoreductase (PFOR), either in the cytosol (G. lamblia) or in hydrogenosomes (T. vaginalis). According to this proposal, the organism leading to G. lamblia subsequently lost the enzymatic components of mitochondrial respiration.

Biochemistry and Metabolism

Axenic Growth of Trophozoites

G. lamblia trophozoites obtained from a rabbit, chinchilla, and cat were first grown axenically (in the absence of exogenous cells) in 1970. HSP-1 medium, a subsequent modification reported in 1976, contained phytone peptone, glucose, l-cysteine HCl, Hanks solution, and human serum This study reported the first human G. lamblia isolate, Portland-1 (P-1). P-1 and WB, an isolate obtained from a symptomatic human who probably acquired his infection in Afghanistan, belong to the same genotype and have been used for many of the studies of G. lamblia. The growth medium has subsequently been modified, and currently the most commonly used medium is modified TYI-S-33. Among the notable requirements for axenic growth are the absolute requirements for a low O_2 concentration, a high cysteine concentration, and the requirement for exogenous lipids, which are obtained from the serum component. When kept at 37 °C, the trophozoites adhere to the glass wall of the tube in which they are grown. This adherence is dependent on glycolysis and on contraction of the proteins of the ventral disk. To date, the other species of Giardia (e.g., G. muris and G. agilis) have not been grown axenically.

Medium for in Vitro Growth of G. Lamblia

Component	Quantity/liter	Final concn
$K_2HPO_4 \cdot H_2O$	1 g	4.4 mM
KH_2PO_4	600 m	4.4 mM
Trypticase (casein peptone)	20 g	
Yeast extract	10 g	
d-Glucose	10 g	56 mM
NaCl	2 g	34 mM
Cysteine-HCl	2 g	16.5 mM
Ascorbic acid	200 mg	1.1 mM
Ferric NH_4-citrate (Sigma F-5879)	22.8 mg	
Bovine bile (Sigma B-8381)	40 ml of 6.5% solution	
NaOH	Bring pH to 7.0–7.2 (1.65 ml of 10 M solution	
Serum (bovine or fetal calf)	100 ml	10%

Formula of TYI-S-33, the most commonly used medium for growth of G. lamblia trophozoites. The ferric-NH4 citrate and bovine bile are generally kept as solutions at 4 °C. All components except the serum are dissolved in water, brought to 200 ml, sterilized on a 0.22- or 0.45-µm-pore-size filter, and added to 700 ml of sterilized water. Heat-inactivated (56 °C for 20 min) serum is then added. During in vitro growth, the organisms are grown in sealed glass containers nearly filled with medium. When this is impossible, such as during cloning by limiting dilution in 96-well plates, a sealed bag containing an anaerobic generator is used. At 4 °C, the shelf life is limited to 5 to 7 days, primarily because of the degradation of the cysteine.

Carbohydrate Metabolism

Most eukaryotic organisms depend primarily on aerobic metabolism for their energy production. However, certain eukaryotes, including Trichomonas spp., Entamoeba spp., and Giardia spp., are characterized by their lack of mitochondria and cytochrome-mediated oxidative phosphorylation. They rely on fermentative metabolism (even when oxygen is present) for energy conservation. Glycolysis and its brief extensions generate ATP, with generation dependent only on substrate level phosphorylation. Glucose is not completely oxidized to CO_2 and H_2O as in aerobic metabolism but is incompletely catabolized to acetate, ethanol, alanine, and CO_2. The balance of end product formation is sensitive to the O_2 tension and glucose concentration in the medium. The proposed pathways of energy production from glucose and aspartate are shown in figure.

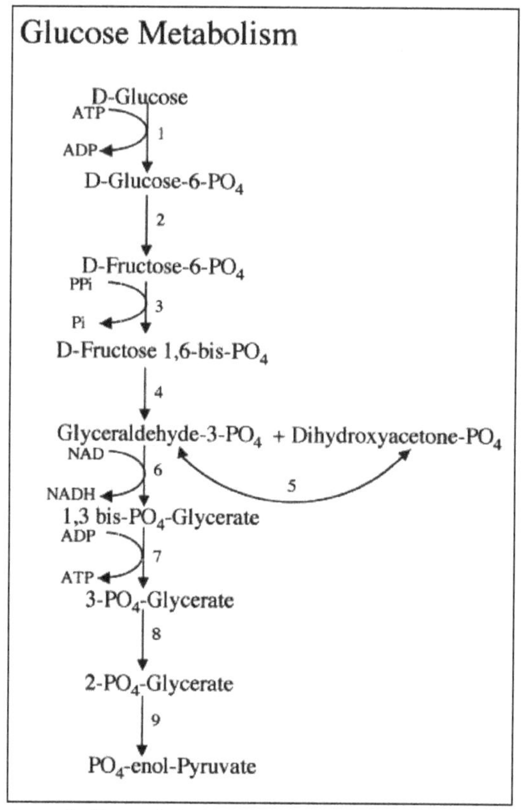

Metabolism of glucose to phosphoenolpyruvate.

Many of the enzymes have been documented in terms of enzymatic activity, isolation of the enzyme, or cloning of the gene encoding the enzyme.

The enzymes are labeled as follows: 1, hexokinase; 2, glucose phosphate isomerase (proposed); 3, pyrophosphate-dependent phosphofructokinase; 4, fructose bisphosphate aldolase; 5, triosephosphate isomerase; 6, glyceraldehyde-3-phosphate dehydrogenase; 7, phosphoglycerate kinase (proposed); 8, phosphoglyceromutase (proposed); 9, enolase (proposed). Certain enzymes (for steps 2, 7, 8, and 9) are suggested on the basis of pathways in other organisms but have not yet been proved for Giardia. This figure has been adapted from material presented in prior reviews and updated from more recent literature as cited for individual enzymes.

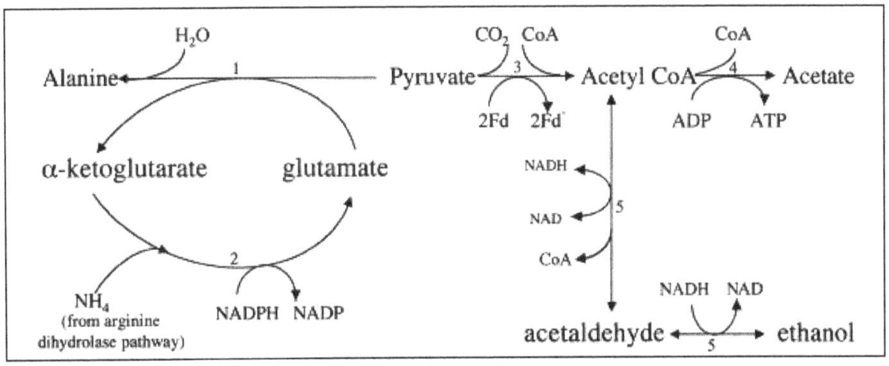

End product synthesis from pyruvate.

The enzymes are labeled as follows: 1, alanine aminotransferase (alanine is produced only under anaerobic conditions); 2, GDH; 3, PFOR (ferredoxin rather than NAD as the electron acceptor); 4, acetyl-CoA synthetase (ADP forming); 5, alcohol dehydrogenase E (ADHE) (has acetaldehyde dehydrogenase activity in the amino terminus which catalyzes the conversion of acetyl-CoA to acetaldehyde and alcohol dehydrogenase activity in the carboxy terminus which catalyzes the conversion of acetaldehyde to ethanol). Acetate is the major product under aerobic conditions; ethanol and alanine are preferentially produced under anaerobic conditions.

The metabolism of trophozoites is markedly affected by small changes in oxygen concentration. Under strictly anaerobic conditions, alanine is the major product of carbohydrate metabolism. Even with the addition of minimal amounts of O_2 (i.e., concentrations of <0.25 μM), ethanol production is stimulated and alanine production is inhibited. With further increases in O_2 concentration, ethanol and alanine production are inhibited. At O_2 concentrations of >46 μM, alanine production is completely inhibited and acetate and CO_2 are the predominant products of energy metabolism. These oxygen concentrations are likely to be relevant to the intestinal milieu in which the trophozoites replicate since the oxygen concentration in this environment is estimated to vary between 0 and 60 μM. Thus, the pathway of metabolism of pyruvate appears to be altered for differing anaerobic or microaerophilic environments.

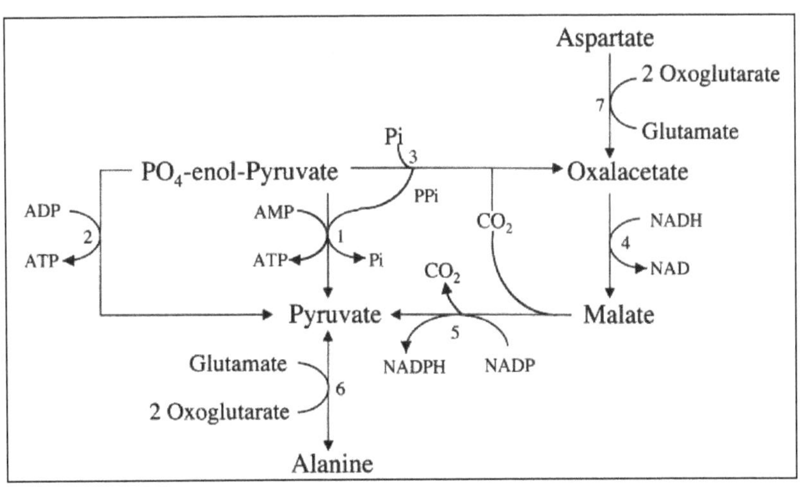

Intermediary pathways to synthesis of pyruvate.

1, Pyruvate phosphate dikinase; 2, pyruvate kinase (the relative roles of pyruvate phosphate dikinase and pyruvate kinase in the conversion of phosphoenolpyruvate to pyruvate have not been determined); 3, phosphoenolpyruvate carboxyphosphotransferase; 4, malate dehydrogenase; 5, malate dehydrogenase (decarboxylating); 6, alanine transaminase; 7, aspartate transaminase. Pi, inorganic phosphate.

Despite the anaerobic metabolism, trophozoites do produce some oxygen free radicals, and a mechanism for detoxification of oxygen is necessary. Detoxification of oxygen free radicals in aerobic organisms is typically accomplished by an enzymatic pathway beginning with superoxide dismutase. Superoxide dismutase activity has been detected by some, but other investigators did not detect superoxide dismutase, catalase, or peroxidase activities in G. lamblia and proposed an H_2O-producing NADH oxidase as the major route of oxygen detoxification in G. lamblia.

Routine Giardia medium contains 50 mM glucose, and many of the metabolic studies of trophozoites have been performed with that medium. When the glucose concentration is reduced to 10 mM, there is little effect on trophozoite growth. At glucose concentrations below 10 mM, replication rates are reduced by nearly 50% and ethanol production is markedly reduced, alanine production is moderately reduced, and acetate production is unaffected. Thus, glucose promotes trophozoite growth but is not absolutely essential. The growth enhancement by glucose was not provided by other sugars.

In T. vaginalis, glycolytic enzymes (glucose to pyruvate) are found in the cytosol but those involved in pyruvate oxidation are localized in a separate organelle, the hydrogenosome. In Giardia, there is no metabolic compartmentalization; rather, all the reactions occur in the cytosol or on the cytosolic surfaces of membranes for PFOR-mediated reactions.

Glucose supplies the major source of energy derived from carbohydrates. Glucose is converted to pyruvate by the Embden-Meyerhof-Parnas and hexose monophosphate shunt pathways. For most eukaryotic and prokaryotic organisms, the conversion of fructose-6-phosphate to fructose-1,6-bisphosphate is an irreversible and regulated step that is catalyzed by an ATP-dependent phosphofructokinase. However, for Giardia, as well as for T. vaginalis and E. histolytica, this reaction is catalyzed by a pyrophosphate-dependent phosphofructokinase. In contrast to the ATP-dependent enzyme, this enzyme catalyzes a reversible reaction and is not a regulated enzyme. The pyrophosphate-dependent phosphofructokinase from Giardia has been cloned and characterized. Triosephosphate isomerase, an enzyme that catalyzes a reversible conversion between dihydroxyacetone phosphate and d-glyceraldehyde-3-phosphate, was cloned by complementation in Escherichia coli and characterized.

Two enzymes that convert phosphoenolpyruvate into pyruvate, ATP-dependent pyruvate kinase, and pyrophosphate-dependent pyruvate phosphate dikinase (PPDK) have been identified. There is a potential energy advantage in the reaction mediated by PPDK, since two molecules of ATP can be generated by a coordinated reaction involving PPDK and adenylate kinase. Adenylate kinase converts two ADP molecules into ATP + ADP by a reaction that is essentially energy neutral, and PPDK converts phosphoenolpyruvate plus AMP into pyruvate + ATP, resulting in the net generation of two ATP molecules during the conversion of phosphoenolpyruvate to pyruvate, rather than the one ATP produced by the pyruvate kinase reaction. Their relative roles in glycolysis have not yet been determined, but the higher specific activity of pyruvate kinase suggests that it may play a major role in glycolysis.

The conversion of pyruvate to acetyl coenzyme A (acetyl-CoA) is catalyzed by PFOR, which utilizes ferredoxin rather than NAD as the electron acceptor, in place of the pyruvate dehydrogenase complex found in aerobic eubacteria and eukaryotes. PFOR is also found in E. histolytica and T. vaginalis.

Metronidazole is an antimicrobial agent with a broad spectrum of activity against anaerobic bacteria and protozoa. It is activated when its 5-nitro group is reduced by ferredoxin that has in turn been reduced by PFOR, generating a toxic nitro radical. Both metronidazole and sodium nitrite, another respiratory inhibitor thought to act by destroying the iron-sulfur center of PFOR, were toxic to G. muris trophozoites, but only sodium nitrite was toxic to cysts. It was proposed that they had different effects because of the inability of metronidazole to enter the cyst, but this proposal was not directly tested. In a study of different G. lamblia isolates, some of which were fully susceptible to metronidazole and some of which were relatively resistant, decreased ferredoxin levels were associated with drug resistance.

Acetyl-CoA can be converted directly to acetate by ADP-forming acetyl-CoA synthetase, resulting in the production of ATP from ADP as acetyl-CoA is converted to acetate. Alternatively, acetyl-CoA is converted to ethanol, using acetaldehyde as an intermediate, by the bifunctional enzyme alcohol dehydrogenase E. Alcohol dehydrogenase E has an acetaldehyde dehydrogenase activity in the amino terminus that catalyzes the conversion of acetyl-CoA to acetaldehyde and an alcohol dehydrogenase activity in the carboxy terminus that converts the acetaldehyde to ethanol.

Amino Acid Metabolism

Amino acids are becoming increasingly recognized as important components of the energy metabolism of G. lamblia. The uptake of aspartate, alanine, and arginine from the extracellular medium, as well as the documentation of glucose-independent metabolism, suggests the potential importance of amino acid metabolism for energy production in Giardia.

The arginine dihydrolase pathway is one potential source of energy. This pathway is present in a number of prokaryotic organisms, but among eukaryotes it has been documented only in T. vaginalis and G. lamblia. In the arginine dihydrolase pathway, arginine is converted to ornithine and ammonia with the generation of ATP from ADP by substrate-level phosphorylation. Ornithine is subsequently exported in exchange for extracellular arginine by a transporter mechanism.

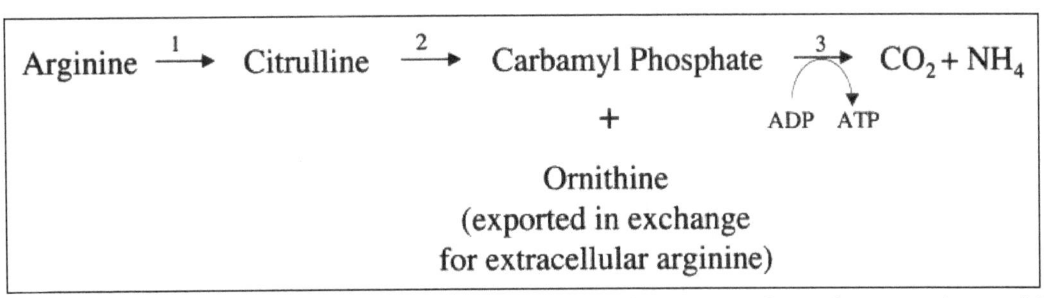

The enzymes are labeled as follows: 1, arginine deiminase; 2, ornithine transcarbamoylase; 3, carbamate kinase.

Aspartate is another potential source of energy. It is converted to oxaloacetate by aspartate transaminase, entering the intermediary pathway, where it is converted to pyruvate via a malate intermediate.

Alanine also appears to play an important role in allowing trophozoites to adapt to hypoosmotic challenge. With an isoosmolar extracellular environment, the intracellular alanine concentration is 50 mM, and with a hypoosmolar challenge, the concentration of alanine rapidly decreases by an active transport mechanism. (In addition to alanine, potassium appears to play a role in osmoregulation of trophozoites.) The secretion of alanine occurs via an alanine transporter that also transports l-serine, glycine and l-threonine, l-glutamine, and l-asparagine. This transporter acts as an antiport, exchanging intracellular alanine for these other amino acids from the extracellular environment.

Aside from the synthesis of alanine as a by-product of energy metabolism, the only other amino acid for which de novo synthesis has been documented is valine. Thus, Giardia lacks synthesis of most amino acids and depends on scavenging them from the intestinal milieu in which the trophozoite replicates.

One of the notable requirements for axenic growth of G. lamblia trophozoites is the absolute requirement for a relatively high concentration of cysteine (16 mM). Cysteine also provides a partial protection from the toxicity of oxygen that is not seen with other reducing agents, including cystine, and therefore appears to be a specific effect of cysteine. Cysteine is not synthesized de novo and is not synthesized from cystine. It appears to be imported into the cell by passive diffusion, although active transport may account for some of the acquisition of cysteine. The importance of free thiol groups on the surfaces of trophozoites was demonstrated by the toxicity of thiol-blocking agents that are unable to penetrate intact cells. This toxicity suggests that these agents are reacting with thiol groups on the trophozoite surfaces, killing the trophozoites. Cysteine appears to be the major thiol group present. When trophozoites are metabolically labeled with radiolabeled cysteine, most of the label is incorporated into the VSPs suggesting that these surface proteins may play a role in protection of the trophozoite from oxygen toxicity.

Lipid Metabolism

The growth of trophozoites predominantly in the duodenum and jejunum initially suggested the possible importance of bile in the growth of Giardia trophozoites. Short-term axenic growth in the absence of serum can be supported by bile. The biliary lipids cholesterol and phosphatidylcholine and the bile salts glycocholate and glycodeoxycholate will also support this growth. Serum is required for longer-term axenic growth, but it has been shown that the Cohn IV-1 fraction of bovine serum (enriched in alpha globulins, lipoproteins, and growth factors) can substitute for whole serum. In fact, insulin-like growth factor II, which is present in fraction IV-1, stimulates trophozoite growth and cysteine uptake. The same fraction from a number of other mammals was also effective in supporting growth, although in some cases antibody depletion was required.

G. lamblia trophozoites do not have the capacity of de novo synthesis of fatty acids, with the possible exception of certain minor fatty acids. However, free fatty acids are toxic to trophozoites, demonstrating 50% lethal doses of 2 to 12 μM. In fact, the toxicity of human milk for G. lamblia trophozoites appears to be mediated through products of milk lipolysis The trophozoites appear to satisfy their lipid requirements by obtaining cholesterol and phosphatidylcholine from the external environment. The cholesterol and phospholipids are supplied by lipoproteins, β-cyclodextrins, and bile salts, with transfer of lipids to the parasite surface being facilitated by bile salts. It has also been suggested that a low level of endocytosis of lipids occurs. Conjugated bile acids appear to be taken up by a carrier-mediated mechanism that includes different carriers for cholyltaurine and cholylglycine.

The major fatty acids found in axenically grown trophozoites are palmitic acid, stearic acid, and oleic acid. Fatty acid desaturase activity, including desaturation of oleate to linoleate and linolenate, has been documented. Arachidonic acid is incorporated into neutral lipids, phospholipids, and a wide variety of cellular lipids, while palmitic acid, myristic acid, and oleic acid are transesterified primarily into phospholipids, including cellular phospholipids, (e.g., phosphatidylglycerol, phosphatidylcholine, phosphatidylethanolamine, and phosphatidylinositol). Interesterification also occurs with incorporation of conjugated fatty acids into phosphatidylglycerol. The toxicity of certain analogs of phosphatidylglycerol for trophozoites has been documented, although the mechanism of this toxicity has not been determined.

Isoprenoids are lipids derived from mevalonate that are commonly found in eukaryotic cells. The most notable end product is cholesterol, but isoprenoids are also incorporated into proteins such as the GTP-binding proteins by posttranslational modification. Isoprenylation of proteins has been demonstrated by the incorporation of radiolabeled mevalonate into trophozoite proteins. Incorporation of mevalonate and cell growth were inhibited in a reversible manner by competitive inhibitors of 3-hydroxy-3-methylglutoryl (HMG)-CoA reductase. Inhibitors of later steps of isoprenylation permanently inhibited cell growth.

Purine and Pyrimidine Salvage

Many of the pathogenic protozoa, including G. lamblia, depend on salvage pathways for obtaining purine nucleosides. In addition, G. lamblia , as well as T. vaginalis and Tritrichomonas foetus, lack pyrimidine synthetic pathways and depends on salvage pathways for obtaining both purine and pyrimidine nucleosides.

Studies of purine metabolism using radiolabeled precursors have demonstrated the incorporation of adenine, adenosine, guanine, and guanosine into nucleotides but no incorporation of components of the de novo synthetic pathway, such as formate, glycine, hypoxanthine, inosine, or xanthine. The likely scenario is that the purine nucleosides (adenosine and guanosine) are imported by a transporter with broad specificity for nucleosides as well as deoxyribonucleosides. The purine nucleosides are then broken down to the bases by their respective hydrolases. Phosphoribosyl 1-pyrophosphate is synthesized by phosphoribosyl 1-pyrophosphate synthetase and reacts with the salvaged purine bases to produce the nucleoside-5′-monophosphate in a reaction catalyzed by the respective monophosphate phosphoribosyltransferase (PRTase). The GPRTases from most eukaryotes utilize hypoxanthine or xanthine as substrate, but the G. lamblia GPRTase is highly specific for guanine, indicating that guanine is the only source for guanine nucleotides. The amino acid sequence of the GPRTase enzyme shows less than 20% identity to the human enzyme. The crystallographic structure of the G. lamblia GPRTase has suggested possible reasons for the unique substrate of the G. lamblia enzyme, such as an aspartate substitution for leucine at position 181.

$$\text{Adenosine} \xrightarrow{1} \text{Adenine} \xrightarrow{2} \text{AMP}$$

$$\text{Guanosine} \xrightarrow{3} \text{Guanine} \xrightarrow{4} \text{GMP}$$

Purine ribonucleoside salvage pathways. The enzymes are labeled as follows: 1, adenosine hydrolase; 2, adenine phosphoribosyltransferase (APRTase); 3, guanosine hydrolase; 4, guanine phosphoribosyltransferase (GPRTase).

G. lamblia trophozoites also depend on the salvage of exogenous thymidine, cytidine, and uridine for the synthesis of the pyrimidine nucleotides Uridine is probably imported by the thymidine transporter and broken down into uracil by a hydrolase or phosphorylase. Cytidine and uridine can also be imported by a broad-specificity transporter, but most cytidine probably enters the synthetic pathway for UTP synthesis rather than being converted more directly to CTP The majority of CTP is probably generated from UTP by CTP synthetase. A low-affinity nucleobase transporter has also been documented. Although the transporter's function has not been determined, it may be involved in export of unusable (e.g., thymine) or excessive quantities of nucleobases.

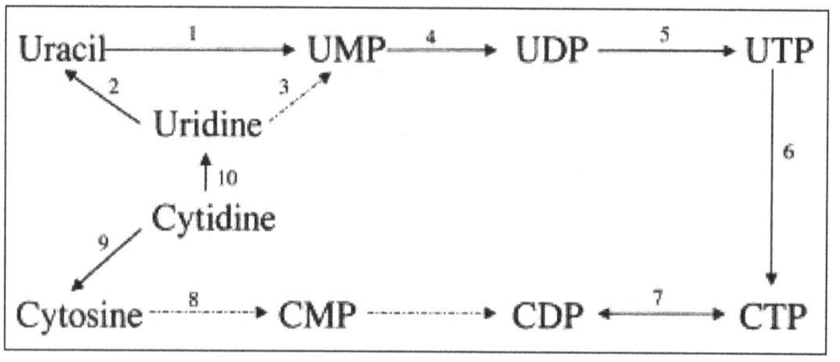

Pyrimidine ribonucleoside salvage pathways.

The enzymes are labeled as follows: 1, uracil phosphoribosyltransferase (UPRTase); 2, uridine/thymine phosphorylase (a uridine hydrolase has not been confirmed); 3, uridine phosphotransferase (kinase) (not confirmed); 4, UMP kinase; 5, UDP kinase; 6, CTP synthetase; 7, CDP kinase; 8, cytosine phosphoribosyltransferase (CPRTase) (low level of activity); 9, cytidine hydrolase; 10, cytidine deaminase.

DNA synthesis also depends on the salvage of exogenous deoxynucleotides. G. lamblia trophozoites lack ribonucleotide reductase and do not incorporate purine bases or nucleosides into DNA. Rather, they depend on the salvage of exogenous purine deoxynucleosides. A single purine deoxynucleoside kinase for the synthesis of deoxynucleotides from their respective deoxynucleosides (deoxyadenosine and deoxyguanosine) and a pyrimidine deoxynucleoside kinase catalyze the production of the pyrimidine deoxynucleotides from the deoxynucleosides thymidine and cytosine.

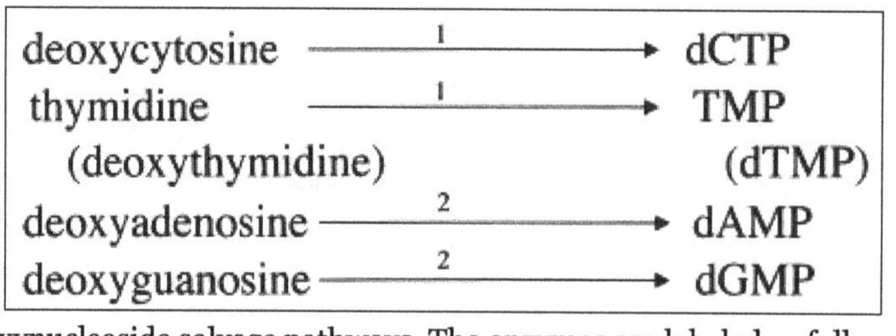

Deoxynucleoside salvage pathways. The enzymes are labeled as follows: 1, pyrimidine deoxynucleoside kinase; 2, purine deoxynucleoside kinase.

Giardia Lamblia Virus

A double-stranded RNA virus was identified in G. lamblia trophozoites following the observation of a 6.2-kb double-stranded RNA contaminating nucleic acid preparations. The virus is nonenveloped and icosahedral, with a diameter of 33 nm, and it infects the nuclei with about 200 copies per nucleus. G. lamblia virus (GLV) has a 100-kDa major capsid protein which depends on a cysteine protease for cleavage into the mature protein. A second open reading frame encodes a 190-kDa

RNA-dependent RNA polymerase. The majority of axenic isolates either harbor or are susceptible to GLV infection (including isolates from genotypes A and B), while a minority of isolates are highly resistant to infection. GLV infection occurs by endocytosis, and susceptibility to infection depends on a specific receptor found of the cell membrane surface. A second virus has also been identified; it also has a 6.2-kb double-stranded RNA genome but encodes a slightly smaller capsid protein (95 kDa), which differs significantly from the 100-kDa capsid protein of GLV.

Surface Antigens and Antigenic Variation

Occurrence and Biological Significance of Antigenic Variation

G. lamblia trophozoites undergo antigenic variation of a family of immunodominant cysteine-rich surface antigens in vitro and in vivo. The initial studies of the surface antigens of G. lamblia showed differences among strains by crossed immunoelectrophoresis and enzyme-linked immunosorbent assay and marked differences in the molecular masses of "excretory-secretory products" from different surface-iodinated G. lamblia isolates. These surface antigens varied in number and in size from approximately 50 to 200 kDa in a study of 19 isolates. A monoclonal antibody (MAb 6E7) for a 170-kDa surface antigen (initially called CRP170 but now called VSPA6) from the WB isolate was cytotoxic for WB trophozoites but not for isolates expressing other surface antigens. WB trophozoites were doubly cloned by limiting dilution and incubated with MAb 6E7; some organisms survived and were totally resistant to the cytotoxic effect of MAb 6E7. One of these cloned lines of organisms (WB1269) had a 68-kDa surface-labeled antigen (initially called CRP68 but now called VSP1269), while another (WB1267) had a 64-kDa surface antigen (VSP1267). A monoclonal antibody (MAb 5C1) reactive with the surface antigen of WB1267 was cytotoxic for WB1267 trophozoites. These trophozoites were incubated with MAb 5C1, and organisms surviving the cytotoxic antibody reacted with neither of the initial MAbs and expressed surface antigens in the 60- to 100-kDa size range. Subsequent data have confirmed that individual organisms express only one VSP at a time the detection of multiple surface-labeled bands in some of the studies resulted from subpopulations of trophozoites expressing several different VSP types. In axenic culture, variation occurs approximately once every 6 to 12 generations for a frequency of 10^{-3} to 10^{-4} .

Antigenic variation has subsequently been confirmed in animal models and in infected human volunteers. Cloned WB trophozoites expressing the VSPA6 were inoculated into gerbils, and trophozoites collected from their intestines 7 days later demonstrated populations of trophozoites expressing multiple VSPs ranging in size from 50 to 170 kDa . Trophozoites collected 28 days after infection were similar to those collected at 7 days. The lack of apparent change from 7 to 28 days argued against a role for acquired immunity in selecting antigenic variants. However, the results were somewhat different in a mouse model of G. lamblia infection. G. lamblia trophozoites in athymic nude mice and in heterozygous *nu/*⁺ mice changed VSP type coincident with the development of a humoral anti-vsp antibody response. In contrast, scid mice did not develop an antibody response and trophozoites did not undergo antigenic variation. Thus, in the mouse model, the antibody response to the VSP may have been the selective force for antigenic variants.

When human volunteers were inoculated with cloned GS trophozoites by duodenal intubation, four of four became infected with trophozoites expressing VSPH7 (72 kDa). Only 1 of 13 became infected with organisms expressing VSPB6 (200 kDa), suggesting that the ability to survive in the human intestine is greater with one VSP than with another. Whether this is due to selection against one VSP or positive selection for the other has not been determined. The subsequent pattern of change during 22 days after infection showed gradual disappearance of VSPH7 followed by the appearance of other VSPs with different immunoreactivity. The loss of VSPH7 was accompanied by the development of serum antibodies to VSPH7, suggesting the possibility that the immune response led to the antigenic variation.

An alternative or additional possible biological reason for antigenic variation might be adaptation to different intestinal environments. Some evidence for this possibility is provided by the difference in protease susceptibility for different VSP antigen types. Trypsin and chymotrypsin were toxic to WB trophozoites expressing VSPA6 (reactive with MAb 6E7). Some organisms survived after exposure to these proteases, and when they were subsequently grown in the presence of trypsin and chymotrypsin, they were no longer susceptible to the cytotoxic effects and expressed alternative VSPs that were not reactive with MAb 6E7.

Antigenic variation has also been documented during the cycle of encystation and excystation. In vitro encystation followed by excystation of trophozoites expressing the VSP TSA417 resulted in the loss of TSA417 from the plasma membrane followed by its appearance in the lysosome-like peripheral vacuoles. During excystation, the TSA417 transcript was replaced by a variety of other vsp transcripts. This occurred too rapidly to be the result of selection of antigenic variants surviving excystation and may have represented an induced antigenic shift. This antigenic variation during excystation could promote survival by evasion of an intestinal immune response or by adaptation to other important intestinal factors.

Thus, the two major hypotheses regarding the purpose of antigenic variation are (i) evasion of the host immune defense and (ii) enabling of the organisms to survive in different intestinal environments. It should be emphasized that these hypotheses are not mutually exclusive. A better understanding of the biological reasons for antigenic variation would be facilitated if the role of the VSPs was known. It would also be informative to know if the free-living diplomonads such as Hexamita have proteins homologous to the VSPs and if they undergo antigenic variation.

Structure and Biochemistry of the VSPs

The initial biochemical studies of the VSPs (then called ES products) indicated that they were protease susceptible. They did not bind to a series of lectins, suggesting that they were not glycosylated. A portion of the gene for one of these surface antigens was subsequently cloned from an expression library using MAb 6E7. Sequence analysis revealed a cysteine-rich (12%) protein with frequent CXXC motifs. Subsequent sequences of a number of vsp genes have shown a number of characteristics common to all vsp genes. They are all cysteine rich (12%), and most cysteines are present in CXXC motifs. When trophozoites are metabolically labeled with radiolabeled cysteine, most of the label is incorporated into the VSPs. It has been suggested that these cysteines are present in the form of disulfide bonds, since free thiol groups have not been detected in purified VSPs. However, the presence of thiol groups on the trophozoite surfaces suggests that the VSP cysteines may not all be disulfide bonded. Whether the level of disulfide bonding changes in vivo with different oxygenation levels of the surrounding environment has not been determined.

After synthesis, the VSPs are transported through the ER to the membrane, where they diffusely coat the membrane. Smaller amounts of VSP can also be detected in the lysosome-like peripheral vacuoles, suggesting the possibility that these vacuoles are involved in recycling of the VSPs. The VSPs have an approximately 14- to 17-amino-acid signal peptide that presumably represents a signal peptide for its transport through the ER. The signal peptide is cleaved, 14 amino acids from VSPH7 of the GS isolate and 17 amino acids from TSA417 of the WB isolate.

The 38 C-terminal amino acids of the VSPs are >90% conserved at the amino acid and nucleotide levels. Each VSP reported to date ends with the amino acid motif CRGKA in both genotype A and B organisms. The conserved C terminus may represent a membrane-anchoring domain. Evidence for this proposal has been reported in work showing a difference in the C termini of membrane-bound and secreted VSP. A 3,500-kDa fragment is cleaved from the C terminus of the membrane-associated VSP prior to secretion. The suggested cleavage site is between the K and S of the highly conserved NKSGLS motif, which comprises the amino acids 33 to 38 from the C terminus. This NKSGLS motif is generally conserved in the genotype A organisms but is not found in the vspH7 gene from GS, a genotype B organism.

The vsp 3′ UTRs have the putative polyadenylation signal common to all G. lamblia strains, AGTPuAAPy, immediately preceded by the sequence ACTPyAGPuT, which sometimes begins within the termination codon.

On the basis of sequences in vsp genes similar to those shown in zinc binding by nuclear proteins, it has been proposed that the VSP genes may encode zinc binding proteins. The ability to bind zinc was shown for VSPH7 and other VSPs but was also found in regions of the VSP that did not contain the "zinc binding" motifs. In addition to zinc, VSPH7 was able to bind other metals including iron. Another laboratory, using VSP4A1 (CRISP90) from a genotype A isolate, demonstrated the ability of reduced but not oxidized VSP to bind zinc. However, the degree of binding was substantially lower than an equimolar zinc:protein amount and was thought to be biologically insignificant.

All of the VPSs reported have potential N-linked glycosylation sites biochemical analysis of VSP4A1 demonstrated O-linked GlcNAc rather than the more commonly found N-linked glycosylation. VSP4A1 was also palmitoylated in the membrane-anchored portion of the C-terminal region. Subsequent work has extended the finding of glycosylation and palmitoylation to other VSPs, including VSPA6 and VSPH7.

Genomic Organization of the VSP Genes and Potential Mechanisms of Antigenic Variation

In addition to their marked 3′ similarity, the vsp genes demonstrate different degrees of identity in other regions. In some cases, there is a high degree of similarity throughout the entire gene. For example, there are two identical convergent copies of the vsp1267 gene approximately 3 kb apart and there is greater than 90% identity in the nonrepeat regions of vspA6 and vspA6-S1 as well as CRP136 and CRP65. These examples suggest that the vsp gene repertoire has been expanded by gene duplication followed by divergence. The sequences for one of these vsp genes (vspA6-S1 or vspG3M) from two geographically separated isolates (WB from Afghanistan and G3M from Peru) were identical over the regions of comparison, suggesting that this divergence is not an extremely rapid process. A number of vsp genes related to TSA417 (vsp417) have been identified from genotype A-1 and A-2 isolates. Comparisons have shown that the differences between genotypes are smaller than the differences between members of this vsp gene family, suggesting that the divergence among the various members of the vsp417 gene family occurred before the divergence between genotypes A-1 and A-2. The vsp genes from the genotype B isolate GS also appear to exist in families but demonstrate similarity to the vsp genes of genotype A only in the conserved 38-amino-acid C terminus and in the CXXC motif.

There are also examples of near identity followed by abrupt divergence, suggesting that in some cases, the vsp repertoire has been expanded by recombination among the vsp genes. For example, vspC5 demonstrates near identity to vspC5-S2 from approximately −150 to +94 and to vspC5-S1 from −150 to +130.

Ascaris Lumbricoides

Ascaris lumbricoides is a nematode, or roundworm, that parasitizes the human gastrointestinal tract. Worldwide, ascariasis is among the most common helminthic human infections with an estimated 800 million to 1.2 billion people infected, and it causes more than 60,000 deaths annually. In fact, ascariasis is considered one of the “neglected tropical diseases,” or NTDs – a diverse group of infectious diseases that exist in tropical and subtropical environments in 149 countries across the world, and which affect more than 1 billion individuals, costing developing economies billions of dollars each year.

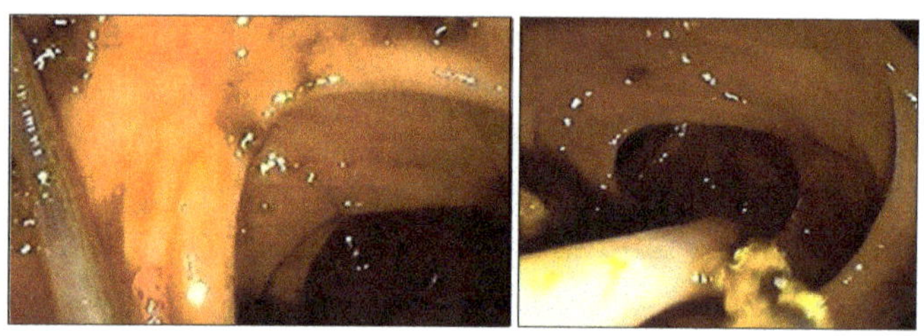

The adult A. lumbricoides worm visualized in the transverse colon during the patient’s screening colonoscopy.

The fact that A. lumbricoides can produce copious numbers of eggs that can survive extreme environmental conditions has made it one of the most prevalent and geographically well-distributed parasitic worms in impoverished socioeconomic regions across the world. In the United States, the occurrence of ascariasis is attributed to immigration from, and travel outside the country to, areas with high prevalence of infection. However, the southeastern region of the United States, due to its temperate, humid climate, has the highest burden of the disease.

The same adult A. lumbricoides worm, approximately 25 cm in length and 6 mm in diameter, after intact removal during the patient’s colonoscopy.

In the life cycle of A. lumbricoides, the definitive host for infection is human and the mode of transmission is the fecal-oral transfer of embryonated eggs. After oral ingestion, via contaminated food or water, the eggs hatch in the small intestine within 4 days and release larvae. The larvae penetrate intestinal walls and enter the portal venous system and lymphatic channels, where they migrate hematogenously and via lymphatics to the liver and then the lungs. There, the larvae move toward and into the alveoli where they mature over a period of 10–14 days, make their way up the tracheobronchial tree to reach the hypopharynx, and are swallowed once more. After returning to the small intestine, these larvae mature into adult worms (females reaching 20–35 cm in length; males reaching 15–30 cm in length) that live in the lumen, typically of the jejunum or ileum. When both female and male adult worms coexist, the females produce fertilized eggs that are passed into the stool. One female adult A. lumbricoides worm inhabiting the small intestine produces approximately 200,000 eggs each day. Under warm, shady, and moist conditions, the eggs can remain viable in soil for up to 10 years and require only 10 to 15 days' time in soil to become infective. Thus, the arrival of larvae back to the small intestine concludes the extraintestinal migration of larvae and begins another molting to become adult worms.

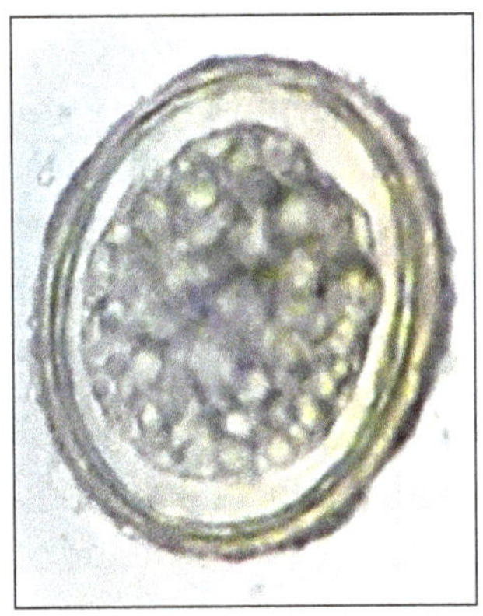

Wet mount of stool from the patient's stool microscopy showing the ovum of A. lumbricoides. ×400 magnification.

Intestinal ascariasis should be suspected in individuals from endemic areas who present with vague nonspecific abdominal symptoms, such as pain, discomfort, dyspepsia, distention, nausea, vomiting, diarrhea, and anorexia, and/or with associated complications, such as intestinal obstruction and hepatobiliary or pancreatic involvement. In endemic regions, patients with high worm loads of A. lumbricoides often present with acute intestinal obstruction. Ochoa studied 311 cases over a period of 25 years from 1963 to 1988. He reported that substantial intestinal infestation, together with the propensity of the worms to interlace and form large masses, often resulted in intestinal obstruction as well as the classic symptoms of abdominal pain, constipation, and vomiting . Intestinal obstruction may also be complicated by appendicitis, intussusception, volvulus, and volvulus with gangrene and perforation. Furthermore, the worms can penetrate biliary and pancreatic ducts, resulting in biliary colic, as well as hepatic and pancreatic dysfunction. The presentation should correlate with relevant epidemiologic exposures, such as to locations with high prevalence of soil-transmitted helminths. The diagnosis of intestinal ascariasis is confirmed via stool microscopy for ova or via detection and examination of adult worms.

Management of the disease and its associated complications consists of anti-helminthic therapy. The mainstays of treatment for ascariasis at this time are albendazole 400 mg orally as a single dose or mebendazole (500 mg orally as a single dose or 100 mg orally twice daily for 3 days). In order to treat pregnant women, pyrantel pamoate 11 mg/kg orally as a single dose, up to a maximum of 1 g, is currently recommended. In certain cases, such as with the complications reported above, surgical and/or endoscopic interventions may be necessary.

Cestoda

Cestoda is a class of parasitic worms in the flatworm phylum (Platyhelminthes). Most of the species - and the best-known - are those in the subclass Eucestoda; they are ribbonlike worms as adults, known as tapeworms. Their bodies consist of many similar units, known as proglottids, which are essentially packages of eggs which are regularly shed into the environment to infect other organisms. Species of the other subclass, Cestodaria, are mainly fish parasites.

All cestodes are parasitic; many have complex life histories, including a stage in a definitive (main) host in which the adults grow and reproduce, often for years, and one or two intermediate stages in which the larvae develop in other hosts. Typically the adults live in the digestive tracts of vertebrates, while the larvae often live in the bodies of other animals, either vertebrates or invertebrates. For example, Diphyllobothrium has at least two intermediate hosts, a crustacean and then one or more freshwater fish; its definitive host is a mammal. Some cestodes are host-specific, while others are parasites of a wide variety of hosts. Some six thousand species have been described; probably all vertebrates can host at least one species.

The adult tapeworm has a scolex, or head, a short neck, and a strobila, or segmented body formed of proglottids. Tapeworms anchor themselves to the inside of the intestine of their host using their scolex, which typically has hooks, suckers, or both. They have no mouth, but absorb nutrients directly from the host's gut. The neck continually produces proglottids, each one containing a reproductive tract; mature proglottids are full of eggs, and fall off to leave the host, either passively in the feces or actively moving. All tapeworms are hermaphrodites, with each individual having both male and female reproductive organs.

Humans are subject to infection by several species of tapeworms if they eat undercooked meat such as pork (Taenia solium), beef (T. saginata), and fish (Diphyllobothrium), or if they live in, or eat food prepared in, conditions of poor hygiene (Hymenolepis or Echinococcus species). The unproven concept of using tapeworms as a slimming aid has been touted since around 1900.

Diversity and Habitat

All 6000 species of Cestoda are parasites, mainly intestinal; their definitive hosts are vertebrates, both terrestrial and marine, while their intermediate hosts include insects, crustaceans, molluscs, and annelids as well as other vertebrates.

T. saginata, the beef tapeworm, can grow up to 20 m (65 ft); the largest species, the whale tapeworm Tetragonoporus calyptocephalus, can grow to over 30 m (100 ft). Species with small hosts tend to be small. For example, vole and lemming tapeworms are only 13–240 mm (0.5–9.4 in) in length, and those parasitizing shrews only 0.8–60 mm (0.03–2.36 in).

Anatomy

Cestodes have no gut or mouth and absorb nutrients from the host's alimentary tract through their specialised neodermal cuticle, or tegument, through which gas exchange also takes place. The tegument also protects the parasite from the host's digestive enzymes and allows it to transfer molecules back to the host.

The body form of adult eucestodes is simple, with a scolex, or grasping head, adapted for attachment to the definitive host, a short neck, and a strobila, or segmented trunk formed of proglottids, which makes up the worm's body. Members of the subclass Cestodaria, the Amphilinidea and Gyrocotylidea, are wormlike but not divided into proglottids. Amphilinids have a muscular proboscis at the front end; Gyrocotylids have a sucker or proboscis which they can pull inside or push outside at the front end, and a holdfast rosette at the posterior end. The Cestodaria have 10 larval hooks while Eucestoda have 6 larval hooks.

Scolex

The scolex, which attaches to the intestine of the definitive host, is often minute in comparison with the proglottids. It is typically a four-sided knob, armed with suckers or hooks or both. In some species, the scolex is dominated by bothria, or "sucking grooves" that function like suction cups. Cyclophyllid cestodes can be identified by the presence of four suckers on their scolices. Other species have ruffled or leaflike scolices, and there may be other structures to aid attachment. In the larval stage the scolex is similarly shaped and is known as the protoscolex.

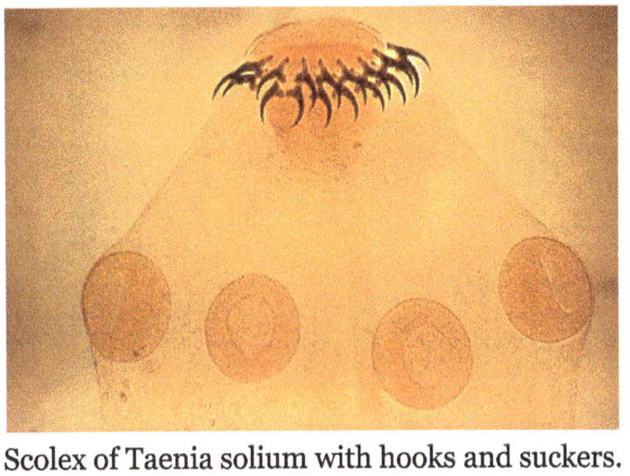

Scolex of Taenia solium with hooks and suckers.

Body Systems

Circular and longitudinal muscles lie under the neodermis, beneath which further longitudinal, dorso-ventral and transverse muscles surround the central parenchyma. Protonephridial cells drain into the parenchyma. There are four longitudinal collection canals, two dorso-lateral and two ventro-lateral, running along the length of the worm, with a transverse canal linking the ventral ones at the posterior of each segment. When the proglottids begin to detach, these canals open to the exterior through the terminal segment.

The main nerve centre of a cestode is a cerebral ganglion in its scolex. Nerves emanate from the ganglion to supply the general body muscular and sensory endings, with two lateral nerve cords running the length of the strobila. The cirrus and vagina are innervated, and sensory endings around the genital pore are more plentiful than in other areas. Sensory function includes both tactoreception (touch) and chemoreception (smell or taste).

Proglottids

Once anchored to the host's intestinal wall, tapeworms absorb nutrients through their surface as their food flows past them. Cestodes are unable to synthesise lipids, which they use for reproduction, and are therefore entirely dependent on their hosts.

The tapeworm body is composed of a series of segments called proglottids. These are produced from the neck by mitotic growth which is followed by transverse constriction. The segments become larger and more mature as they are displaced backwards by newer segments. Each proglottid contains an independent reproductive tract, and like some other flatworms, cestodes use flame cells (protonephridia), located in the proglottids, for excretion. The sum of the proglottids is called a strobila, which is thin and resembles a strip of tape; from this is derived the common name "tapeworm". Proglottids are continually being produced by the neck region of the scolex, as long as the scolex is attached and alive.

Mature proglottids are essentially bags of eggs, each of which is infective to the proper intermediate host. They are released and leave the host in faeces, or migrate outwards as independent motile proglottids. The number of proglottids forming the tapeworm ranges from three to four thousand. Their layout comes in two forms: craspedote, meaning any given proglottid is overlapped by the previous proglottid, or acraspedote, indicating the proglottids do not overlap.

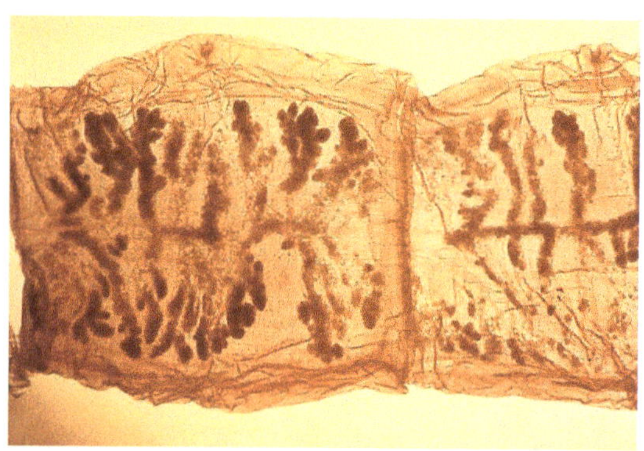

Two proglottids of Taenia solium. This species has 7 to 13 branches of the uterus on each side (above and below in this micrograph).

Reproduction

Cestodes are exclusively hermaphrodites, with both male and female reproductive systems in each body. The reproductive system includes one or more testes, cirri, vas deferens, and seminal vesicles as male organs, and a single lobed or unlobed ovary with the connecting oviduct and uterus as female organs. The common external opening for both male and female reproductive systems is known as the genital pore, which is situated at the surface opening of the cup-shaped atrium. Though they are sexually hermaphroditic and cross-fertilization is the norm, self-fertilization sometimes occurs and makes possible the reproduction of a worm when it is the only individual in its host's gut. During copulation, the cirri of one individual connect with those of the other through the genital pore, and then spermatozoa are exchanged.

Life Cycle

Cestodes are parasites of vertebrates, with each species infecting a single definitive host or group of closely related host species. All but amphilinids and gyrocotylids (which burrow through the gut or body wall to reach the coelom) are intestinal, though some life-cycle stages rest in muscle or other tissues. The definitive host is always a vertebrate but in nearly all cases, one or more intermediate hosts are involved in the lifecycle, typically arthropods or other vertebrates. Infections can be long-lasting; in humans, tapeworm infection may last as much as 30 years. No asexual phases occur in the lifecycle, as they do in other flatworms, but the lifecycle pattern has been a crucial criterion for assessing evolution among Platyhelminthes.

Cestodes produce large numbers of eggs, but each one has a low probability of finding a host. To increase their chances, different species have adopted various strategies of egg release. In the Pseudophyllidea, many eggs are released in the brief period when their aquatic intermediate hosts are abundant (semelparity). In contrast, in the terrestrial Cyclophyllidea, proglottids are released steadily over a period of years, or as long as their host lives (interoparity). Another strategy is to have very long-lived larvae; for example, in Echinococcus, the hydatid larvae can survive for ten years or more in humans and other vertebrate hosts, giving the tapeworm an exceptionally long time window in which to find another host.

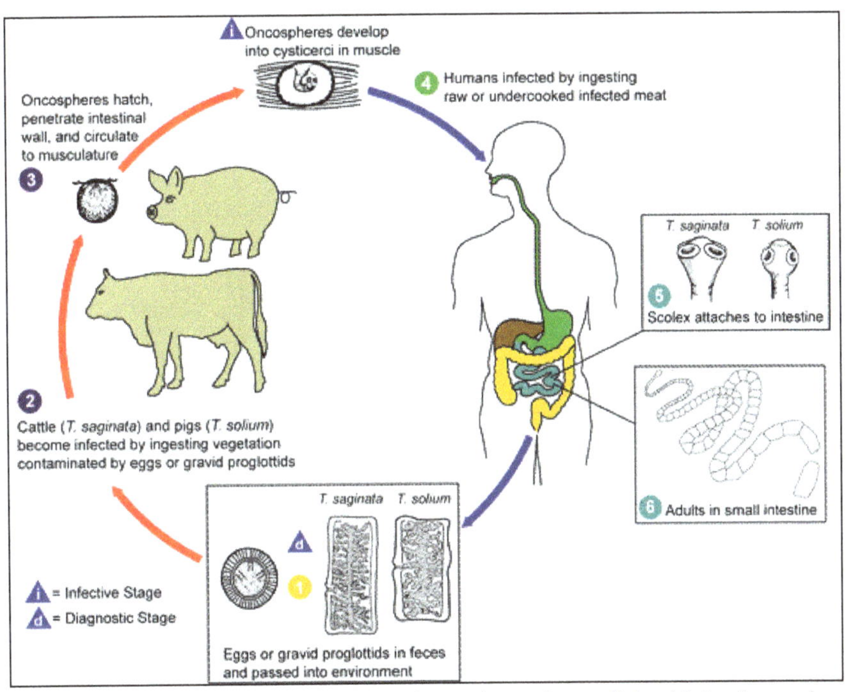

Life cycle of the eucestode Taenia: Inset 5 shows the scolex, a disk with hooks on the end. Inset 6 shows the tapeworm's whole body, in which the scolex is the tiny, round tip in the top left corner, and a mature proglottid has just detached.

Many tapeworms have a two-phase lifecycle with two types of host. The adult Taenia saginata lives in the gut of a primate such as a human, its definitive host. Proglottids leave the body through the anus and fall to the ground, where they may be eaten with grass by a grazing animal such as a cow. This animal then becomes an intermediate host, the oncosphere boring through the gut wall and migrating to another part of the body such as the muscle. Here it encysts, forming a cysticercus.

The parasite completes its lifecycle when the intermediate host passes on the parasite to the definitive host, usually when the definitive host eats contaminated parts of the intermediate host, for example a human eating raw or undercooked meat. Another two-phase lifecycle is exhibited by Anoplocephala perfoliata, the definitive host being an equine and the intermediate host an oribatid mite.

Diphyllobothrium exhibits a more complex, three-phase lifecycle. If the eggs are laid in water, they develop into free-swimming oncosphere larvae. After ingestion by a suitable freshwater crustacean such as a copepod, the first intermediate host, they develop into procercoid larvae. When the copepod is eaten by a suitable second intermediate host, typically a minnow or other small freshwater fish, the procercoid larvae migrate into the fish's flesh where they develop into plerocercoid larvae. These are the infective stages for the mammalian definitive host. If the small fish is eaten by a predatory fish, its muscles too can become infected.

Schistocephalus solidus is another three-phase example. The intermediate hosts are copepods and small fish, and the definitive hosts are waterbirds. This species has been used to demonstrate that cross-fertilisation produces a higher infective success rate than self-fertilisation.

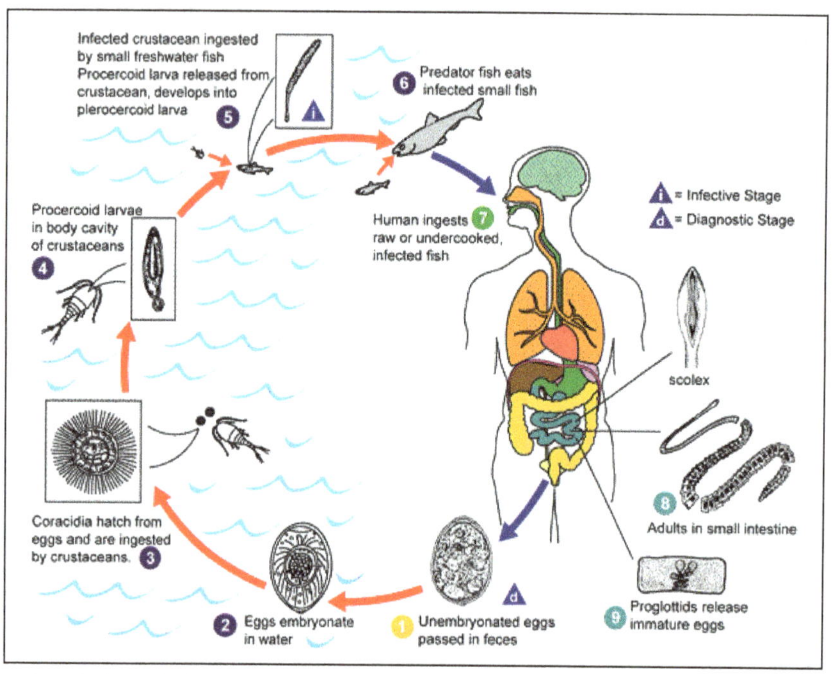

Life cycle of Diphyllobothrium latum relies on at least three hosts, crustaceans, fish, and humans. Other fish-eating mammals like bears can equally serve as definitive hosts.

Host Immunity

Hosts can become immune to infection by a cestode if the lining, the mucosa, of the gut is damaged. This exposes the host's immune system to cestode antigens, enabling the host to mount an antibody defence. Host antibodies can kill or limit cestode infection by damaging their digestive enzymes, which reduces their ability to feed and therefore to grow and to reproduce; by binding to their bodies; and by neutralising toxins that they produce. When cestodes feed passively in the gut, they do not provoke an antibody reaction.

Eucestoda

Eucestoda is the larger of the two subclasses of flatworms in the class Cestoda (the other subclass is Cestodaria). The Eucestoda are commonly referred to as tapeworms. Larvae have six posterior hooks on the scolex (head), in contrast to the ten-hooked Cestodaria. All tapeworms are endoparasites of vertebrates, living in the digestive tract or related ducts. Examples are the pork tapeworm (Taenia solium) with a human definitive host, and pigs as the secondary host, and Moniezia expansa, the definitive hosts of which are ruminants.

Body Structure

Adult Eucestoda have a white-opaque dorso-ventrally flattened appearance, and are elongated, ranging in length from a few mm to 25 m. Almost all members, except members of the orders Caryophyllidea and Spathebothriidea, are polyzoic with repeated sets of reproductive organs down the body length, and almost all members, except members of the order Dioecocestidae, are protandral hermaphrodites. Most except caryophyllideans consist of a few to 4000 proglottids (segments) that show a characteristic body differentiation pattern into scolex (head), neck, and strobila.

The scolex, located at the anterior end, is a small (usually less than 1 mm) holdfast organ with specific systems for fastening itself to materials: rostrum, acetabula, suckers, bothria, grooves, and hooks. The small neck region, directly behind the scolex, consists of an undifferentiated tissue region of proglottid proliferation, leading into a zone of increasing and continuous proglottid differentiation. As such, the main and largest section of the body, the strobila, consists of a chain of increasingly mature proglottids. These cytological processes are not well understood at present.

Members of the Eucestoda have no mouth or digestive tract, and instead absorb nutrients through a layer of microtriches over the tegument at the shared body wall surface. In addition to the body wall, several other systems are common to the whole length of the tapeworm, including excretory canals, nerve fibers, and longitudinal muscles. The excretory system is responsible for osmoregulation and consists of blind-ending flame bulbs communicating through a duct system. The nervous system, often referred to as a "ladder system," is a system of longitudinal connectives and transverse ring commissures.

Reproduction

The reproductive systems develop progressively along the differentiated proglottids of the strobila region, with each proglottid developing one or two sets of sexual organs that differentiate at different times in a species-specific pattern, usually male-first. Thus, moving in the posterior direction of the continuously maturing proglottid chain, there are proglottids with mature male reproductive organs, then proglottids with mature female reproductive organs, and then proglottids with fertilized eggs in the uterus, a condition commonly referred to as "gravid."

Proglottids

An atrium on the lateral margin of each proglottid contains the openings to both the male and female genital ducts. Follicular testes produce sperm, which are carried by a system of ducts to the cirrus, an eversible copulatory organ that usually has a hypodermic system of spines and a holdfast system of hooks. The main specialized female reproductive organs are an ovary that produces eggs and a vitellarium that produces yolk cells. Yolk cells travel in a duct system to the oviduct, where, in a modified region, the ovum is enclosed in a shell with yolk cells. After the gonads and their ducts have finished maturing, the female reproductive organs begin to mature. The oviduct develops a vagina and enlarges into the uterus, where fertilization and embryonic development occur.

Egg formation is a result of copulation. A proglottid can copulate with itself, with other proglottids in the same worm, or with proglottids in other worms, and hypodermic fertilization sometimes occurs. When a gravid proglottid that is distended with an embryo reaches the end of the strobila, it detaches and passes out of the host intact with feces, with or without some tissue degeneration. In the order Pseudophyllidea, the uterus has a pore and the proglottid sheds the shelled embryo, only becoming detached when exhausted.

Some members of the Eucestoda (such as Echinococcus, Sparganum, Taenia multiceps sp., and Mesocestoides sp.) can reproduce asexually through budding, which initiates a metagenesis of alternating sexually and asexually reproducing generations.

Life Stages

A tapeworm can live from a few days to over 20 years. Eucestoda ontogenesis continues through metamorphosing in different larval stages inside different hosts. The initial six-hooked embryo, known as an oncosphere or hexacanth, forms through cleavage. In the order Pseudophyllidea, it remains enclosed in a ciliated embryophore. The embryo continues to develop in other host species, with two intermediate hosts generally needed. It gains entry to its first intermediate host by being eaten.

Except for members of the order Taeniidae, the first intermediate host is an arthropod, and except for in the case of Archigetes spp. (which can attain sexual maturity in freshwater oligochaeta), the second host is usually a fish, but can be another invertebrate or vertebrate. After the scolex has differentiated and matured in the larval stage, growth will stop until a vertebrate eats the intermediate host, and then the strobila develops. Adult tapeworms often have a high final host specificity, with some species only found in one host vertebrate.

Medical Importance

Cysticercosis and Neurocysticercosis

Humans become final hosts of various species of Taenia by eating raw or undercooked beef (usually T. saginata) or pork (usually T. solium). A human becomes an abnormal intermediate host through swallowing or antiperistaltic contractions during regurgitation carrying eggs or gravid proglottids to the stomach. At this point, larvae hatch when exposed to enzymes and penetrate the intestinal wall, travelling through the body through blood vessels to tissues like the brain, the eye, muscles, and the nervous system (called neurocysticercosis).

At these sites, the parasites lodge and form cysts, a condition called cysticercosis, producing inflammatory reactions and clinical issues when they die, sometimes causing serious or fatal damage. In the eye, the parasites can cause visual loss, and infection of the spine and adjacent leptomeninges can cause paresthesias, pain, or paralysis.

Neurocysticercosis can cause headaches, seizures, vomiting, visual disturbances, behavioral abnormalities, focal neurological deficits, and signs of increased intracranial pressure such as papilledema and altered alertness. Hydrocephalus can occur, and larvae can threaten the host's life by obstructing ventricles in the subarachnoid space.

Echinococcosis (Hydatid Disease)

Humans become accidental hosts to worms of the genus Echinococcus, playing no role in the worm's biological cycle. This can result in echinococcosis, also called hydatid disease. Humans (usually children) become infected by direct contact with dogs and eating food contaminated with dog feces. Common sites of infection are the liver, the lungs, muscles, bones, kidneys, and the spleen.

Eggs hatch in the gastrointestinal tract after the consumption of contaminated food, after which the larvae travel to the liver through portal circulation. Here, the larvae are trapped and usually develop into hydatid cysts. While the liver is the first filter for trapping them, the lungs act as the second filter site, trapping most of the larvae that are not trapped by the liver. Some larvae escape from the lungs to cause cysts in other tissues.

When a larva becomes established in tissue, it develops into a "bladderworm" or "hydatid" and can cause various cancer-like cysts that may rupture and interact with nearby organs. Most cases are asymptomatic, and the mortality rate is low, but various complications from these interactions may lead to debilitating illness.

Hymenolepiasis

Arthropods are intermediate hosts of Hymenoleptis nana, otherwise known as the "dwarf tapeworm," while humans are used as final hosts. Humans become infected and develop hymenolepiasis through eating infected arthropods, ingesting eggs in water inhabited by arthropods, or from dirty hands. This is a common and widespread intestinal worm.

While light infections are usually asymptomatic, autoinfection through eating the eggs of worms in the intestines is possible, and it can lead to hyperinfection. Humans can also become hyperinfected through ingesting grain products contaminated by infected insects. Infections involving more than two thousand worms can cause many different gastrointestinal symptoms and allergic responses. Common symptoms include chronic urticaria, skin eruption, and phlyctenular keratoconjunctivitis.

Diphyllobothriasis

Diphyllobothriasis is caused by infection with Diphyllobothrium latum (also known as the "broad tapeworm" or "fish tapeworm") and related species. Humans become infected by eating raw, undercooked, or marinated fish acting as a second intermediate or paratenic host harboring metacestodes or plerocercoid larvae.

Clinical symptoms are due to the large size of the tapeworm, which often reaches a length exceeding 15 m (49 ft). The most common symptom is pernicious anemia, caused by the absorption of vitamin B12 by the worm. Other symptoms include various intestinal issues, slight leukocytosis, and eosinophilia.

Sparganosis

Sparganosis is caused by the plerocercoid larvae of the tapeworm Spirometra. Humans become infected by drinking contaminated water, eating raw or poorly cooked infected flesh, or from using poultices of raw infected flesh (usually raw pork or snake) on skin or mucous membranes.

The most common symptom is a painful, slowly growing nodule in the subcutaneous tissues, which may migrate. Infection in the eye area can cause pain, irritation, edema, and excess watering. When the orbital tissues become infected, the swelling can cause blindness. An infected bowel may become perforated. Brain infection can cause granulomas, hematomas, and abscesses.

Leech

Leeches are segmented parasitic or predatory worms that belong to the phylum Annelida and comprise the subclass Hirudinea. They are closely related to the oligochaetes, which include the earthworms, and like them have soft, muscular, segmented bodies that can lengthen and contract. Both groups are hermaphrodites and have a clitellum, but leeches typically differ from the oligochaetes in having suckers at both ends and in having external annulations that do not correspond with their internal segmentation. The body is relatively solid, and the spacious body cavity found in other annelids, the coelom, is reduced to small channels.

The majority of leeches live in freshwater environments, while some species can be found in terrestrial and marine environments. The best-known, such as the medicinal leech, Hirudo medicinalis, are hematophagous, attaching themselves to a host with a sucker and feeding on blood, having first secreted the peptide hirudin to prevent the blood from clotting. A minority of leech species are predatory, mostly preying on small invertebrates.

In aquatic species, the eggs are enclosed in a cocoon which is usually attached to something solid, but terrestrial species often conceal the cocoon under a log or in a crevice. Almost 700 species of leech are currently recognised, of which some 100 are marine, 90 terrestrial and the remainder freshwater.

Leeches were used in medicine from ancient times until the 19th century to draw blood from patients. In modern times, leeches find medical use in treatment of joint diseases such as epicondylitis and osteoarthritis, extremity vein diseases, and microsurgery, while hirudin is a valuable drug for some blood-clotting disorders.

Diversity and Phylogeny

Some 700 species of leech have been described, of which some 100 are marine, 90 terrestrial and the remainder freshwater. The name for the subclass, Hirudinea, comes from the Latin hirudo, hirudinis, a leech.

The more primitive Acanthobdellidea are often included with the leeches, but some authors treat them as a separate clitellate group. The Branchiobdellida include cleaning symbionts such as Xironodrilus appalachius, which lives on the crayfish Cambarus chaugaensis. Both groups live in freshwater.

True leeches of the infraclass Euhirudinea have both anterior and posterior suckers. They are divided into two groups: Arhynchobdellida and Rhynchobdellida.

"Rhynchobdellida" (paraphyletic) are jawless leeches, armed with a muscular, straw-like proboscis puncturing organ in a retractable sheath. The Rhynchobdellae consist of two families:

- Glossiphoniidae are flattened freshwater leeches, mostly parasitic on vertebrates such as turtles, and unique among annelids in carrying their young under their bodies.
- Piscicolida are marine or freshwater ectoparasites of fish, with cylindrical bodies and usually well-marked, bell-shaped, anterior suckers.

Arhynchobdellida lack a proboscis and may or may not have jaws armed with teeth. Arhynchobellids are divided into two orders:

- Hirudiniformes are jawed leeches, armed with teeth, including the blood-feeding European medical leech, Hirudo medicinalis. Within this order, the Hirudidae are aquatic and the Haemadipsidae are terrestrial and mainly blood-feeders, including Haemadipsa sylvestris, the Indian leech and Haemadipsa zeylanica, the Japanese mountain leech.
- Erpobdelliformes: The freshwater or amphibious worm-leeches are carnivorous and equipped with a relatively large, toothless mouth to ingest insect larvae, molluscs, and other annelid worms, which are swallowed whole. The Pharyngobdella have six to eight pairs of eyes; Gnathobdella has five pairs. The Erpobdellidae live in freshwater habitats.

The phylogenetic tree of the leeches and their annelid relatives is based on molecular analysis of mitochondrial cytochrome c oxidase. Both the former classes "Polychaeta" (bristly marine worms) and "Oligochaeta" (including the earthworms) are paraphyletic.

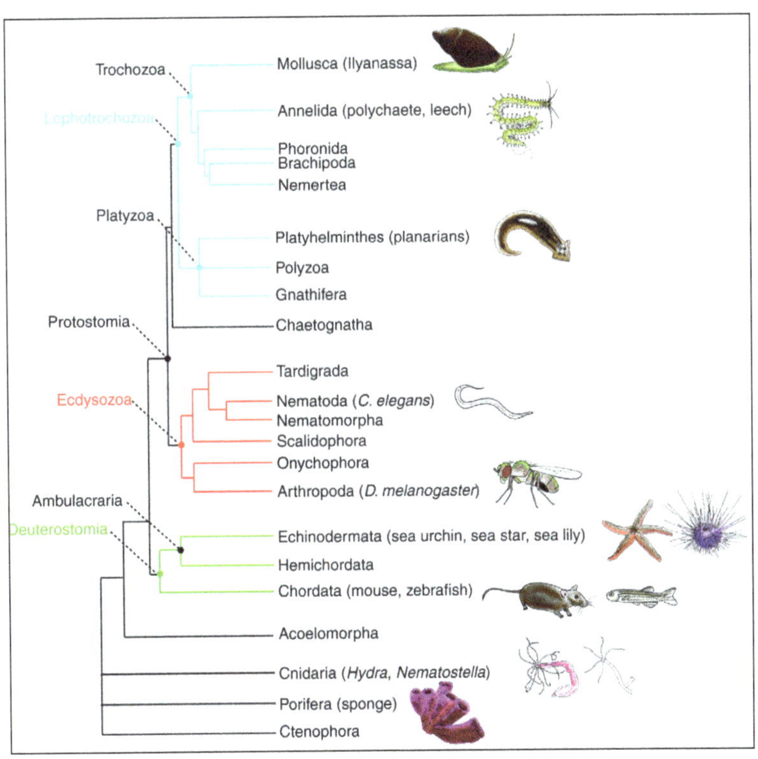

Anatomy and Physiology

Leeches show a remarkable similarity to each other in morphology, very different from typical annelids which are basically cylindrical, with a fluid-filled space, the coelom (body cavity). In leeches, the coelom is reduced to four slender longitudinal channels, and the interior of the body is filled with a solid dermis in between the various organs. Typically, the body is dorso-ventrally flattened and tapers at both ends. Longitudinal and circular muscles in the body wall are supplemented by diagonal muscles, giving the leech the ability to adopt a large range of body shapes and show great flexibility. Most leeches have a sucker at both the anterior (front) and posterior (back) ends, but some primitive leeches have a single sucker at the back.

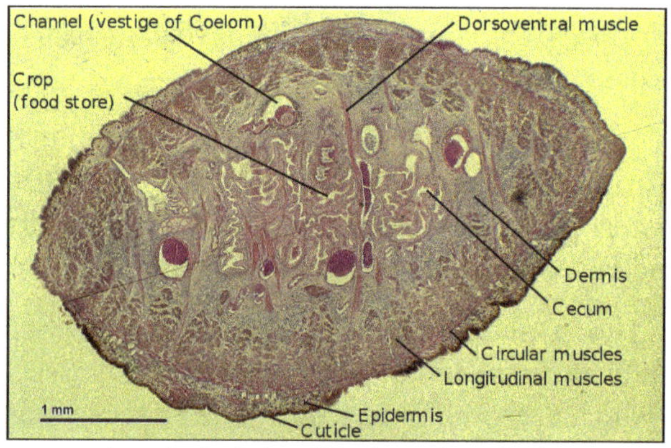

Leech anatomy in cross-section: the body is solid, the coelom (body cavity) reduced to channels, with circular, longitudinal, and transverse muscles making the animal strong and flexible.

Like other annelids, the leech is a segmented animal, but unlike other annelids, the segmentation is masked by the external annulation (ring markings), with the body surface being divided into 102 annuli, whereas its internal structure consists of 32 segments. Of these segments, the first five are designated as the head and include the anterior brain, several ocelli (eyespots) dorsally and the sucker ventrally. The following 21 mid-body segments each contain a nerve ganglion, and between them contain two reproductive organs, a single female gonopore and nine pairs of testes. The last seven segments contain the posterior brain and are fused to form the animal's tail sucker.

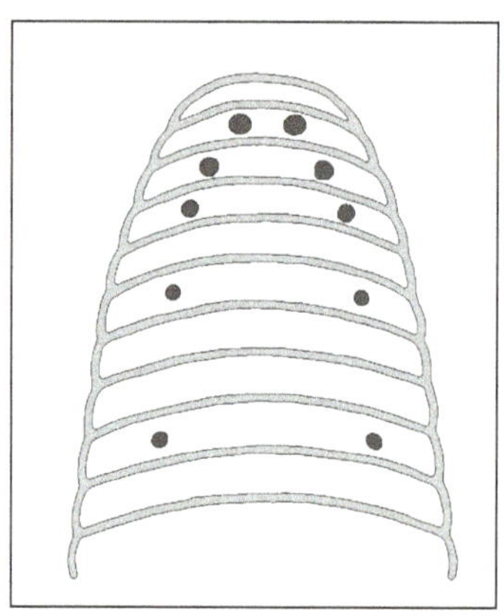

The number and position of eyes are essential for distinguishing the leech species.

The body wall consists of a cuticle, an epidermis and a thick layer of fibrous connective tissue in which are embedded the circular muscles, the diagonal muscles and the powerful longitudinal muscles. There are also dorso-ventral muscles. The coelomic channels run the full length of the body, the two main ones being on either side. The coelom has taken over the function of the hemal system (blood vessels) in other annelids. Part of the lining epithelium consists of chloragogen cells which are used for the storage of nutrients and in excretion. There are ten to seventeen pairs of metanephridia (excretory organs) in the mid-region of the leech. From these, ducts typically lead to a urinary bladder, which empties to the outside at a nephridiopore.

Reproduction and Development

Leeches are protandric hermaphrodites, with the male reproductive organs, the testes, maturing first and the ovaries later. In hirudinids, a pair will line up with the clittelar regions in contact, with the anterior end of one leech pointing towards the posterior end of the other; this results in the male gonopore of one leech being in contact with the female gonopore of the other. The penis passes a spermatophore into the female gonopore and sperm is transferred, and probably stored in, the vagina.

Some jawless leeches (Rhynchobdellida) and proboscisless leeches (Arhynchobdellida) lack a penis, and in these, sperm is passed from one individual to another by hypodermic injection. The leeches intertwine and grasp each other with their suckers. A spermatophore is pushed by one through the integument of the other, usually into the clitellar region. The sperm is liberated and passes to the ovisacs, either through the coelomic channels or interstitially through specialist "target tissue" pathways.

Some time after copulation, the small, relatively yolkless eggs are laid. In most species, an albumin-filled cocoon is secreted by the clitellum and receives one or more eggs as it passes over the female gonopore. The cocoon is fixed to a submerged object, or in the case of terrestrial leeches, deposited under a stone or buried in damp soil. The cocoon of Hemibdella soleae is attached to a suitable fish host, and that of a branchiobdellid, to the shell of a crayfish on which the young will become ectosymbionts. The glossiphoniids brood their eggs, either by attaching the cocoon to the substrate and covering it with their ventral surface, or by securing the cocoon to their ventral surface, and even carrying the newly hatched young to their first meal.

When breeding, most marine leeches leave their hosts and become free-living in estuaries. Here they produce their cocoons, after which the adults of most species die. When the eggs hatch, the juveniles seek out potential hosts when these approach the shore. Leeches mostly have an annual or biannual life cycle.

Feeding and Digestion

About three quarters of leech species feed on blood extracted from a host while the remainder are predators. Leeches either have a protusible pharynx, commonly called a proboscis, or a non-protusible pharynx which may or may not be armed with jaws.

In the proboscisless leeches, the jaws of Arhynchobdellids are at the front of the mouth, and have three blades set at an angle to each other. In feeding these slice their way through the skin of the host, leaving a Y-shaped incision. Behind the blades is the mouth, located ventrally at the anterior end of the body. It leads successively into the pharynx, a short oesophagus, a crop (in some species), a stomach and a hindgut, which ends at an anus located just above the posterior sucker. The stomach may be a simple tube, but the crop, when present, is an enlarged part of the midgut with a number of pairs of ceca that stores ingested blood. The leech secretes an anticoagulant, hirudin, in its saliva which prevents the blood from clotting before ingestion. A mature medicinal leech may feed only twice a year, taking months to digest a blood meal.

The bodies of predatory leeches are similar, though instead of jaws many have a protrusible proboscis, which for most of the time they keep retracted into the mouth. Such leeches are often ambush predators that lie in wait till they can strike prey with the proboscises in a spear-like fashion. Predatory leeches feed on small invertebrates such as snails, earthworms and insect larvae. The prey is usually sucked in and swallowed whole. Some Rhynchobdellida, however, suck the soft tissues from their prey, making them intermediate between predators and blood-suckers.

When they are hungry, leeches use their anterior suckers to connect to hosts for feeding. Once attached, leeches use a combination of mucus and suction to stay in place while they inject hirudin into the hosts' blood streams. In general, sanguivorous leeches are non host-specific, and do little harm to their host, dropping off after consuming a blood meal. Some marine species, however, remain attached until it is time to reproduce. If present in great numbers on a host, these can be debilitating, and in extreme cases, cause death.

Leeches are unusual in that they do not produce amylases, lipases or endopeptidases. This lack of endopeptidases means the mechanism of protein digestion cannot follow the same sequence as it would in all other animals in which endopeptidases first split proteins into peptides, and the exopeptidases then degrade the peptides. However, they do produce intestinal exopeptidases which remove amino acids from the long protein molecules one by one, possibly aided by proteases from endosymbiotic bacteria in the intestine. This evolutionary choice of exopeptic digestion in Hirudinea distinguishes these carnivorous clitellates from oligochaetes, and may explain why digestion in leeches is so slow.

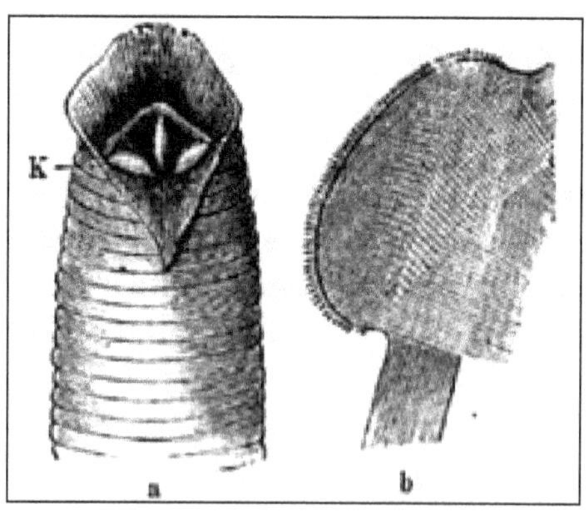

Mouthparts and sucker.

Deficiency of digestive enzymes and of B complex vitamins is compensated for by enzymes and vitamins produced by endosymbiotic microflora. In Hirudo medicinalis, these supplementary factors are produced by an obligatory symbiotic relationship with two bacterial species, Aeromonas veronii and a still-uncharacterised Rikenella species. Nonbloodsucking leeches, such as Erpobdella punctata, are host to three bacterial symbionts, Pseudomonas, Aeromonas, and Klebsiella spp. (a slime producer). The bacteria are passed from parent to offspring in the cocoon as it is formed.

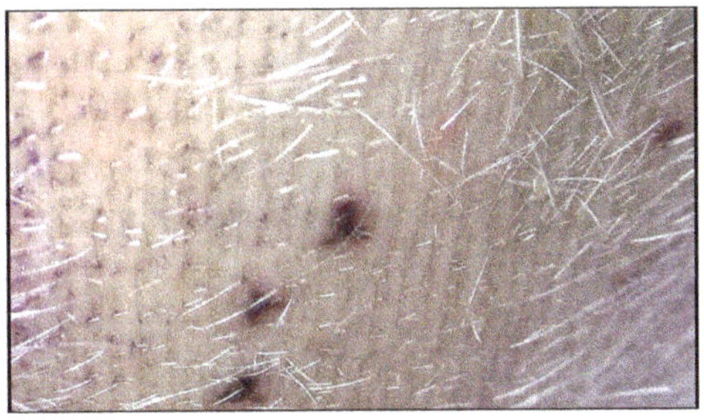

Leech bites on a cow's udder.

Leech attacking a slug.

Nervous System

The nervous system of leeches is formed of a small number of large nerve cells; this has resulted in leeches being used as model organisms for the study of invertebrate nervous systems. The main nerve centre consists of the cerebral ganglion above the gut and another ganglion beneath it, with connecting nerves forming a ring around the pharynx a little way behind the mouth. A nerve cord runs backwards from this in the ventral coelomic channel, with 21 pairs of ganglia in segments 6 to 26. In segments 27 to 33, other paired ganglia fuse to form the caudal ganglion. Several sensory nerves connect directly to the cerebral ganglion; there are sensory and motor nerve cells connected to the ventral nerve cord ganglia in each segment.

Leeches have between two and ten pigment spot ocelli, arranged in pairs towards the front of the body. There are also sensory papillae arranged in a lateral row in one annulation of each segment. Each papilla contains many sensory cells. Some rhynchobdellids have the ability to change colour dramatically by moving pigment in chromatophore cells; this process is under the control of the nervous system but its function is unclear as the change in hue seems unrelated to the colour of the surroundings.

Leeches can detect touch, vibration, movement of nearby objects, and chemicals secreted by their hosts; freshwater leeches crawl or swim towards a potential host standing in their pond within a few seconds. Species that feed on warm-blooded hosts move towards warmer objects. Many leeches avoid light, though some blood feeders move towards light when they are ready to feed, presumably increasing the chances of finding a host.

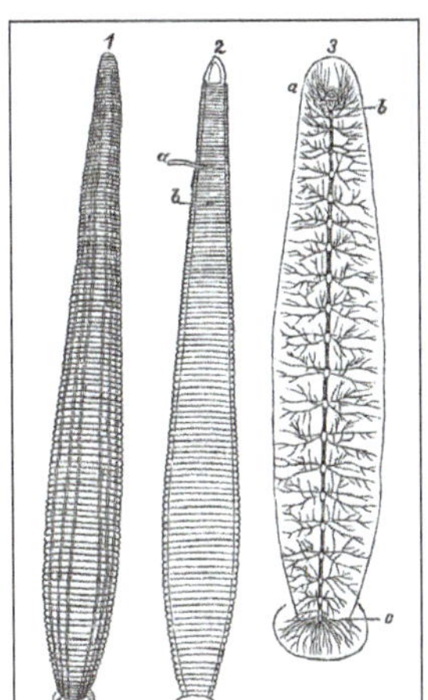

The leech and its nervous system.

Gas Exchange

Leeches live in damp surroundings and in general respire through their body wall. The exception to this is in the Piscicolidae, where branching or leaf-like lateral outgrowths from the body wall form gills. Some rhynchobdellid leeches have an extracellular haemoglobin pigment, but this only provides for about half of the leech's oxygen transportation needs.

Movement

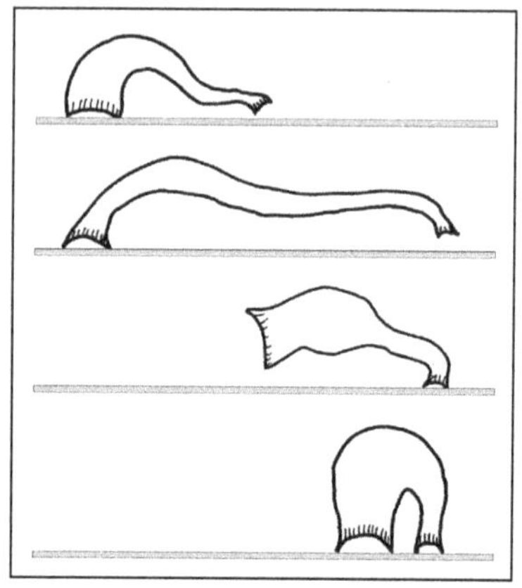

Leech moving by looping using its front and back suckers.

Leeches move using their longitudinal and circular muscles, which other annelids such as earthworms use for peristalsis, with the addition of the use of their posterior and anterior suckers (one on each end of the body) to enable them to progress by looping or inching along, in the manner of geometer moth caterpillars. The posterior end is attached to the substrate, and the anterior end is projected forward peristaltically by the circular muscles until it touches down, as far as it can reach, and the anterior end is attached. Then the posterior end is released, pulled forward by the longitudinal muscles, and reattached; then the anterior end is released, and the cycle repeats. Leeches explore their environment with head movements and body waving.

References

- Chang, M. S. (2002), "Operational issues in the control of the vectors of Brugia", Annals of Tropical Medicine and Parasitology, 96 (2): S71–S76, doi:10.1179/000349802125002392, PMID 12625920
- Ancylostoma-duodenale-habitat-morphology-and-life-cycle, aschelminthes, animals-2: biologydiscussion.com, Retrieved 9 January, 2019
- Junghanss, Jeremy Farrar, Peter J. Hotez, Thomas (2013). Manson's Tropical Diseases: Expert Consult - Online(23rd ed.). Oxford: Elsevier/Saunders. Pp. E49–e52. ISBN 9780702053061
- Symptoms-causes, pinworm, syc-20376382, diseases-conditions: mayoclinic.org, Retrieved 10 February, 2019
- Valentine, Genevieve (30 October 2013). "Medical Magic Leads To Terror In 'Parasite'". National Public Radio. Retrieved 15 June 2018
- Whipworm-infection-causes, health: healthline.com, Retrieved 11 March, 2019
- Melrose WD (2002). "Lymphatic filariasis: New insights into an old disease". Int J Parasitol. 32 (8): 947–60. Doi:10.1016/S0020-7519(02)00062-0. PMID 12076624
- Filarial-worm, animal : britannica.com, Retrieved 12 April, 2019

- Kim, Bong Jin, et al. "Heavy Hymenoleptis nana Infection Possibly Through Organic Food: Report of a Case." The Korean Journal of Parasitology. 2014. P. 85–87
- Mansonella-ozzardi, accounts: animaldiversity.org, Retrieved 13 May, 2019
- Thomson, MC; Obsomer, V; Dunne, M; Connor, SJ; Molyneux, DH (2000). "Satellite mapping of Loa loa prevalence in relation to ivermectin use in west and central Africa". The Lancet. 356 (9235): 1077–1078. Doi:10.1016/s0140-6736(00)02733-1

5

Parasitic Arthropods

The arthropods are the invertebrate animal having an exoskeleton, paired jointed appendages and a segmented body. Some of the common parasitic arthropods are fleas, demodex and ticks. The topics elaborated in this chapter will help in gaining a better perspective about these types of arthropod parasites.

Arthropods form a huge assemblage of small coelomate animals with "jointed limbs" (hence the name arthro-pods). They exhibit segmentation of their bodies (metamerism) which is often masked in adults because their 10-25 body segments are combined into 2-3 functional groups (called tagmata). They exhibit varying degrees of cephalization whereby neural elements, sensory receptors and feeding structures are concentrated in the head region. Arthropods possess a rigid cuticular exoskeleton consisting mainly of tanned proteins and chitin. The exoskeleton is usually hard, insoluble, virtually indigestible and impregnated with calcium salts or covered with wax. The exoskeleton provides physical and physiological protection and serves as a place for muscle attachment. Skeletal plates are joined by flexible articular membranes and the joints are hinges or pivots made from chondyles and sockets.

Biodiversity

The main arthropod assemblages include crustaceans (crabs, lobsters, crayfish, shrimp), arachnids (spiders, scorpions, ticks, mites) and insects (beetles, bugs, earwigs, ants, bees, termites, butterflies, moths, crickets, roaches, fleas, flies, mosquitoes, lice). Most parasitic arthropods belong to 2 main classes: the 6-legged insects, and the 8-legged arachnids.

- Insects have 3 distinct body parts, commonly called the head, thorax and abdomen. The head has 2 antennae and the thorax has 6 legs arranged in 3 bilateral pairs. Many insect species also have 2 pairs of wings attached to the thorax. Parasitic insect species include fleas, flies and lice which actively feed on host tissues and fluids at some stage in their life-cycles.

- Arachnids have 2 body parts known as the prosoma (or cephalothorax) and opisthosoma (or abdomen). The cephalothorax has 8 legs arranged in 4 bilateral pairs and arachnids do not have wings or antennae. Important parasitic assemblages include the ticks and mites which bite into tissues and feed off host fluids.

Collectively, arthropods account for a substantial share of global biodiversity, both in terms of species richness and relative abundance. There are over 1,000,000 species of insects and over 50,000 species of arachnids. They are very successful and adaptable organisms and are capable of forming large populations due to their rapid and fertile reproduction rates. Many species are also able to withstand adverse environmental conditions by undergoing periods of developmental arrest (diapause). The protection afforded by their exoskeletons allows them to colonize many habitats and they overcome the problem of growing larger in a non-expandable exoskeleton by undergoing periodic moulting (or ecdysis) which is mediated by hormones. Developmental stages between moults are referred to as instars. Moulting is a complex process and its timing is mediated by many environmental and physiological cues. It involves detachment of the hypodermis from the procuticle, partial resorption of the old cuticle, production of a new epicuticle, dehiscence (splitting) of the old cuticle, emergence of the animal, stretching and expansion of the new cuticle by air and/or water intake, and then sclerotization of the new cuticle.

Life-cycles

Adult arthropods are generally small in size, most are visible but some remain microscopic. Arthropod sexes are separate and fertilization is internal. A wide range of mating behaviours, insemination and egg production strategies are involved. In most species, the egg develops into a larva: i.e. a life-cycle stage that is structurally distinct from the adult and must undergo metamorphosis (structural reorganization) before becoming an adult. This metamorphosis may be complete (involving major changes during a pupation stage) or incomplete (involving gradual changes in nymph stages). For example, the grub-like larval stages of flies and fleas form cocoon-like pupae where they undergo complete metamorphosis and emerge as radically-different adult insects. In contrast, the larval instars (or nymphs) of lice, ticks and mites undergo incomplete metamorphosis through a series of moults gradually becoming more adult-like in appearance.

	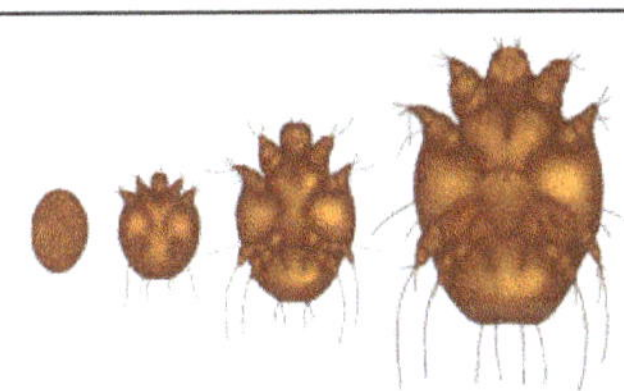
complete metamorphosis	incomplete metamorphosis
hatch moult hatch	hatch moult moult
egg----------larva----------pupa----------adult	egg----------larva----------nymph----------adult

Arthropods are involved in nearly every kind of parasitic relationship, either as parasites themselves or as hosts/vectors for other micro-organisms (including viruses, bacteria, protozoa and helminths). They are generally ectoparasitic on, or in, the skin of vertebrate hosts. Many species are haematophagous (suck blood) while others are histophagous (tissue-feeders) and bite or burrow in dermal tissues causing trauma, inflammation and hypersensitivity reactions. Infestations are transmitted from host-to-host either by direct contact or by free-living larvae or adults actively seeking hosts.

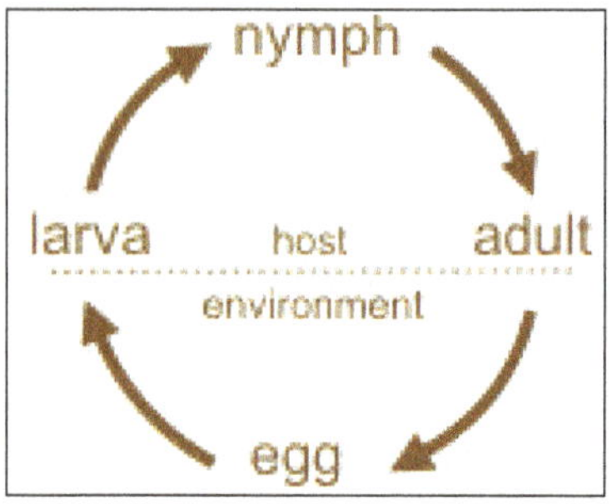

Direct

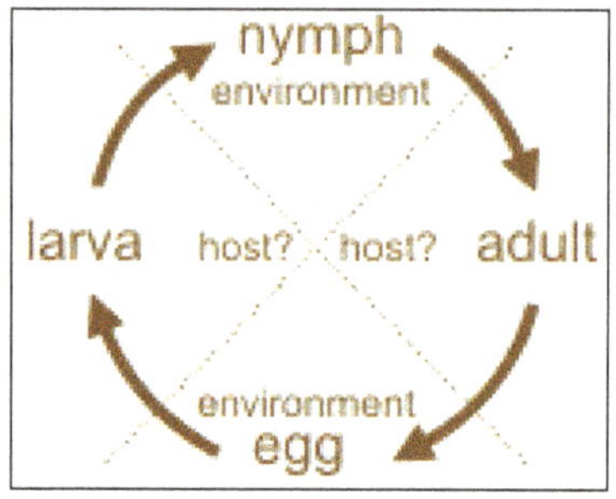

Host-seeking (larva or adult parasitic.)

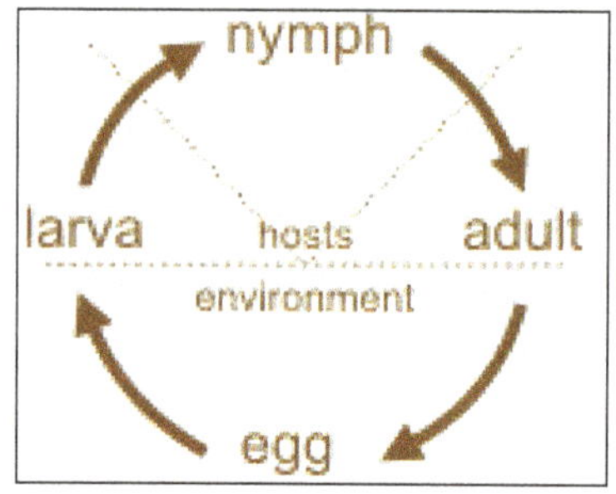

Host-seeking (all feeding stages parasitic.)

- Direct transmission of infective stages occurs when hosts come into close contact with each other or share quarters, bedding or clothing. Larvae, nymphs or adults may cross from one host to another, while eggs or pupae may contaminate shared environments. Insects (fleas and lice) and arachnids (mites) rely on close contact between hosts.
- Many adult insects actively seek hosts in order to feed or lay eggs. Winged insects (mosquitoes, flies) fly to new hosts to feed while fleas jump onto passing hosts. Some adult flies (botflies) do not feed on their hosts but deposit eggs from which larvae emerge and feed on host tissues and exudates.
- Tick larvae actively seek hosts by climbing vegetation and questing for passing hosts. Some species complete their life-cycle on the same host (one-host ticks) while others detach after feeding and drop to the ground to moult before seeking new hosts as nymphs or adults (two-host or three-host ticks).

Taxonomy

Insects exhibit extraordinary biodiversity, both in terms of species richness (numbers of species) and relative abundance (population sizes). Most parasitic species belong to three main groups: the jumping fleas (Siphonaptera); the winged flies (Diptera); and the wingless lice (Phthiraptera).

- Fleas Are bilaterally-flattened wingless insects with enlarged hindlimbs specially adapted for jumping (up to 100 times their body length). Jumping feats are accomplished using elastic resilin pads which expand explosively when uncocked from the compressed state. Fleas undergo complete metamorphosis whereby grub-like larvae form pupae from which adult fleas emerge. The larvae are not parasitic but feed on debris associated mainly with bedding, den or nest material, whereas the adult stages are parasitic and feed on host blood. There are some 2,500 flea species, most parasitic on mammals (especially rodents) and some on birds. They vary in the time spent on their hosts ranging from transient feeders (rodent fleas) to permanent attachment (sticktight fleas and burrowing chigoes).
- Flies and mosquitoes are winged insects with two pairs of wings attached to the thorax and a well-developed head with sensory and feeding organs. They undergo complete

metamorphosis involving a pupation stage. Different species vary in their feeding habits, both as adults (parasitic or free-living) and larvae (parasitic or free-living). There are over 120,000 species belonging to 140 families. Two main suborders are recognized on the basis of structural differences, Nematocera (adult stages parasitic, larval stages often free-swimming) and Brachycera (adult stages parasitic or free-living, larvae stages often predaceous).

- Lice Are small wingless insects, dorsoventrally flattened, with reduced or no eyes and enlarged tarsal claws for clinging. All lice undergo gradual metamorphosis and there are no free-living stages. Eggs are cemented to hair/feathers whereas nymphs and adults cling to hair/feathers. Two orders of lice are recognized on the basis of their mouthparts: the Mallophaga (chewing/biting lice) with some 3,000 species infesting birds and mammals; and the Anoplura (sucking lice) with 500 species found on mammals.

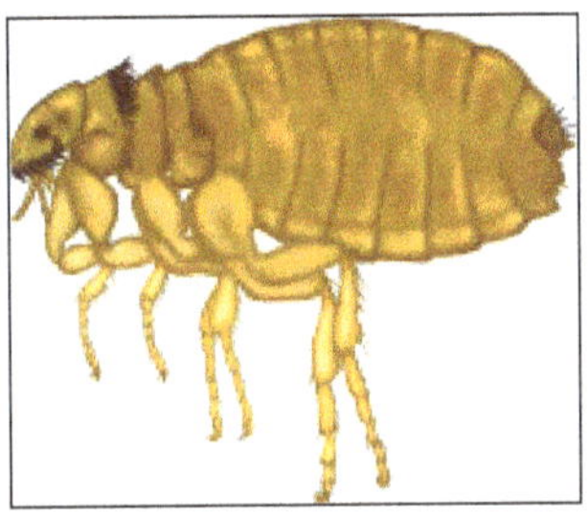

Fleas

Flies

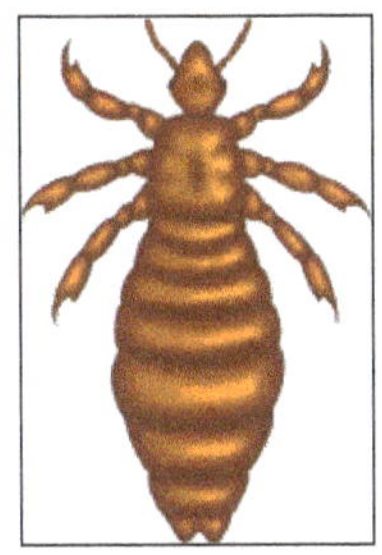

Lice

Many non-spider arachnids (subclass Acari) are found as parasites on animal or plant hosts. They belong to two main groups: the macroscopic ticks and the microscopic mites. Many species are important in human and animal medicine as causes of disease or as transmission vectors for other pathogens.

- Ticks are epidermal parasites of terrestrial vertebrates that may cause anaemia, dermatosis, paralysis, otoacariasis and other infections (transmit viral, bacterial, rickettsial, spirochaete, protozoal and helminth pathogens). They feed mainly on blood and their mouthparts are armed with small backward-facing teeth to aid in attachment. All ticks undergo gradual/incomplete metamorphosis whereby larval and nymphal instars resemble adults. The integument is relatively thick and respiration occurs via spiracles (usually only one pair) and trachea. Two major families of ticks are recognized on the basis of many morphological features: the Ixodidae (hard ticks with a tough cuticle and a large anterodorsal scutum) with some 650 species that infest mammals, birds and reptiles; and the Argasidae (soft ticks with a leathery integument and no scutum) with 160 species that infest mainly birds and some mammals.

- Mites are microscopic arachnids which undergo gradual or incomplete metamorphosis. Adults and nymphs have 4 pairs of legs whereas larvae have 3 pairs. Over 30,000 species of mites have been described, many are free-living species, some are plant parasites while others are parasitic on terrestrial and aquatic hosts. Most parasitic species feed on skin debris or suck lymph, some burrow into the skin, some live in hair follicles, and some in the ear canals. Their mouthparts are variable in form but the hypostome is never armed with teeth. The integument is usually thin and three orders are recognized on the basis of their respiratory systems: the Mesostigmata with respiratory spiracles (stigmata) near the third coxae; the Prostigmata (Trombidiformes) with spiracles between the chelicerae or on the dorsal hysterosoma; and the Astigmata (Sarcoptiformes) without tracheal systems as they respire through the tegument.

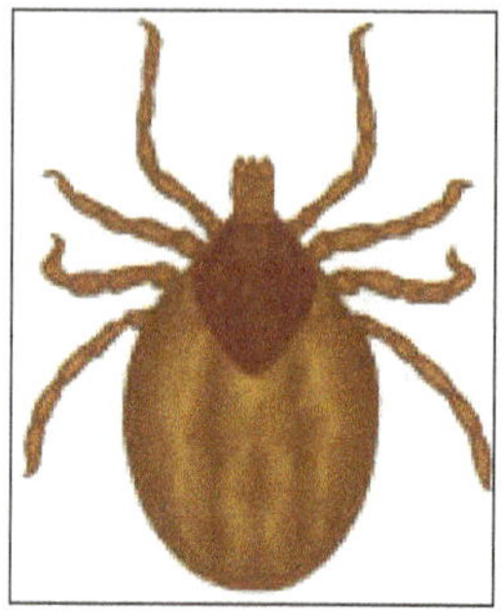

Ticks

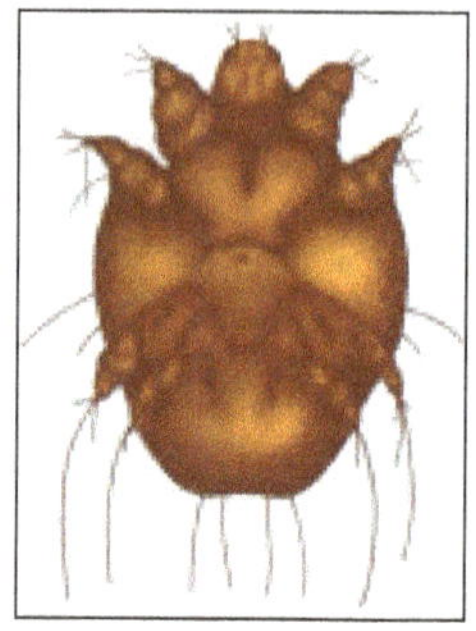

Mites

Flea

Flea, the common name for the order Siphonaptera, includes 2,500 species of small flightless insects that survive as external parasites of mammals and birds. Fleas live by consuming blood or hematophagy, from their hosts. Adult fleas grow to about 3 mm or .12 in long, are usually brown, and have bodies that are "flattened" sideways, or narrow, enabling them to move through their host's fur or feathers. They lack wings, but have strong claws preventing them from being dislodged; mouthparts adapted for piercing skin and sucking blood, and hind legs extremely well adapted for jumping. They are able to leap a distance of some 50 times their body length, a feat second only to jumps made by another group of insects, the superfamily of froghoppers. Fleas' larvae are worm-like with no limbs; they have chewing mouthparts and feed on organic debris left on their host's skin.

The Siphonaptera are most closely related to the snow scorpionflies, or snow fleas in the UK, formally the Boreidae, placing them within the Endopterygote insect order Mecoptera. Fleas arose in the early Cretaceous, most likely as ectoparasites of mammals, before moving on to other groups including birds. Each species of flea is more or less a specialist with respect to its host animal species: many species never breed on any other host, though some are less selective. Some families of fleas are exclusive to a single host group; for example, the Malacopsyllidae are found only on armadillos, the Ischnopsyllidae only on bats, and the Chimaeropsyllidae only on elephant shrews.

The oriental rat flea, Xenopsylla cheopis, is a vector of Yersinia pestis, the bacterium which causes bubonic plague. The disease was spread by rodents such as the black rat, which were bitten by fleas that then infected humans. Major outbreaks included the Plague of Justinian, c. 540 and the Black Death, c. 1350, both of which killed a sizeable fraction of the world's population.

Fleas appear in human culture in such diverse forms as flea circuses, poems like John Donne's erotic The Flea, works of music such as by Modest Mussorgsky, and a film by Charlie Chaplin.

Morphology and Behaviour

Fleas are wingless insects, 1/16 to 1/8-inch (1.5 to 3.3 mm) long, that are agile, usually dark colored (for example, the reddish-brown of the cat flea), with a proboscis, or stylet, adapted to feeding by piercing the skin and sucking their host's blood through their epipharynx. Flea legs end in strong claws that are adapted to grasp a host.

Unlike other insects, fleas do not possess compound eyes but instead only have simple eyespots with a single biconvex lens; some species lack eyes altogether. Their bodies are laterally compressed, permitting easy movement through the hairs or feathers on the host's body (or in the case of humans, under clothing). The flea body is covered with hard plates called sclerites. These sclerites are covered with many hairs and short spines directed backward, which also assist its movements on the host. The tough body is able to withstand great pressure, likely an adaptation to survive attempts to eliminate them by scratching.

Fleas lay tiny, white, oval eggs. The larvae are small and pale, have bristles covering their worm-like bodies, lack eyes, and have mouth parts adapted to chewing. The larvae feed on organic matter, especially the feces of mature fleas, which contain dried blood. Adults feed only on fresh blood.

Jumping

Their legs are long, the hind pair well adapted for jumping; a flea can jump vertically up to 7 in (18 cm) and horizontally up to 13 in (33 cm), making the flea one of the best jumpers of all known animals (relative to body size), second only to the froghopper. The flea jump is so rapid and forceful that it exceeds the capabilities of muscle, and instead of relying on direct muscle power, fleas store muscle energy in a pad of the elastic protein named resilin before releasing it rapidly (like a human using a bow and arrow). Immediately before the jump, muscles contract and deform the resilin pad, slowly storing energy which can then be released extremely rapidly to power leg extension for propulsion. To prevent premature release of energy or motions of the leg, the flea employs a "catch mechanism". Early in the jump, the tendon of the primary jumping muscle passes slightly behind the coxa-trochanter joint, generating a torque which holds the joint closed with the leg close to the body. To trigger jumping, another muscle pulls the tendon forward until it passes the joint axis, generating the opposite torque to extend the leg and power the jump by release of stored energy. The actual take off has been shown by high-speed video to be from the tibiae and tarsi rather than from the trochantera (knees).

Life Cycle and Development

Fleas are holometabolous insects, going through the four lifecycle stages of egg, larva, pupa, and imago (adult). In most species, neither female nor male fleas are fully mature when they first emerge but must feed on blood before they become capable of reproduction. The first blood meal triggers the maturation of the ovaries in females and the dissolution of the testicular plug in males, and copulation soon follows. Some species breed all year round while others synchronise their activities with their hosts' life cycles or with local environmental factors and climatic conditions. Flea populations consist of roughly 50% eggs, 35% larvae, 10% pupae, and 5% adults.

Egg Stage

The number of eggs laid depends on species, with batch sizes ranging from two to several dozen.

The total number of eggs produced in a female's lifetime (fecundity) varies from around one hundred to several thousand. In some species, the flea lives in the nest or burrow and the eggs are deposited on the substrate, but in others, the eggs are laid on the host itself and can easily fall off onto the ground. Because of this, areas where the host rests and sleeps become one of the primary habitats of eggs and developing larvae. The eggs take around two days to two weeks to hatch. Experiments have shown that fleas lay more eggs on hosts which have limited food intakes, and that eggs and larvae survive better under these conditions, perhaps because the host's immune system is compromised.

Larva Stage

Flea larvae emerge from the eggs to feed on any available organic material such as dead insects, faeces, conspecific eggs, and vegetable matter. In laboratory studies, some dietary diversity seems necessary for proper larval development. Blood-only diets allow only 12% of larvae to mature, whereas blood and yeast or dog chow diets allow almost all larvae to mature. Another study also showed that 90% of larvae matured into adults when the diet included nonviable eggs. They are blind and avoid sunlight, keeping to dark, humid places such as sand or soil, cracks and crevices, under carpets and in bedding.

Pupa Stage

Given an adequate supply of food, larvae pupate and weave silken cocoons after three larval stages. Within the cocoon, the larva moults for a final time and undergoes metamorphosis into the adult form. This can take just four days, but may take much longer under adverse conditions, and there follows a variable-length stage during which the pre-emergent adult awaits a suitable opportunity to emerge. Trigger factors for emergence include vibrations (including sound), heat (in warm-blooded hosts), and increased levels of carbon dioxide, all of which stimuli may indicate the presence of a suitable host. Large numbers of pre-emergent fleas may be present in otherwise flea-free environments, and the introduction of a suitable host may trigger a mass emergence.

Adult Stage

Once the flea reaches adulthood, its primary goal is to find blood and then to reproduce. Female fleas can lay 5000 or more eggs over their life, permitting rapid increase in numbers. Generally speaking, an adult flea only lives for 2 or 3 months. Without a host to provide a blood meal, a flea's life can be as short as a few days. Under ideal conditions of temperature, food supply, and humidity, adult fleas can live for up to a year and a half. Completely developed adult fleas can live for several months without eating, so long as they do not emerge from their puparia. Optimum temperatures for the flea's life cycle are 21 °C to 30 °C (70 °F to 85 °F) and optimum humidity is 70%.

Adult female rabbit fleas, Spilopsyllus cuniculi, can detect the changing levels of cortisol and corticosterone hormones in the rabbit's blood that indicate it is getting close to giving birth. This triggers sexual maturity in the fleas and they start producing eggs. As soon as the baby rabbits are born, the fleas make their way down to them and once on board they start feeding, mating, and laying eggs. After 12 days, the adult fleas make their way back to the mother. They complete this mini-migration every time she gives birth.

Dog flea (from top) larva, egg, pupa and adult.

Flea larva.

Taxonomy and Phylogeny

Between 1735 and 1758, the Swedish naturalist Carl Linnaeus first classified insects, doing so on the basis of their wing structure. One of the seven orders into which he divided them was "Aptera", meaning wingless, a group in which as well as fleas, he included spiders, woodlice and myriapods. It wasn't until 1810 that the French zoologist Pierre André Latreille reclassified the insects on the basis of their mouthparts as well as their wings, splitting Aptera into Thysanura (silverfish), Anoplura (sucking lice) and Siphonaptera (fleas), at the same time separating off the arachnids and crustaceans into their own subphyla.

Fleas are related to the Diptera (true flies) and the Mecoptera (scorpion flies) as shown in the cladogram, based on a 2008 analysis of four loci (18S and 28S ribosomal DNA, cytochrome oxidase II, and elongation factor 1-alpha) for 128 flea taxa from around the world. The Boreidae (snow scorpionflies) are the sister clade to the Siphonaptera.

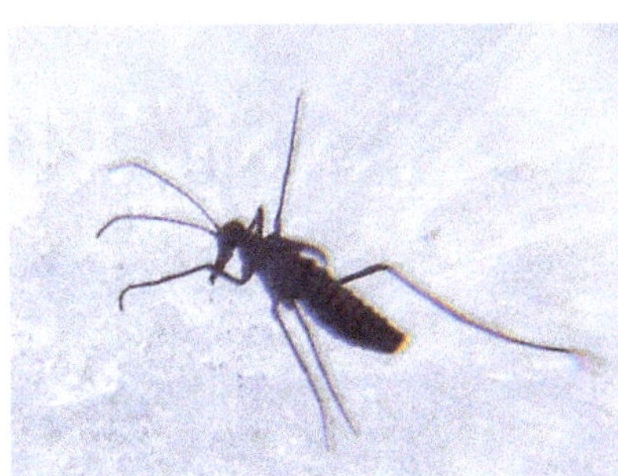

The snow scorpionflies (Boreidae) are the sister clade to the Siphonaptera.

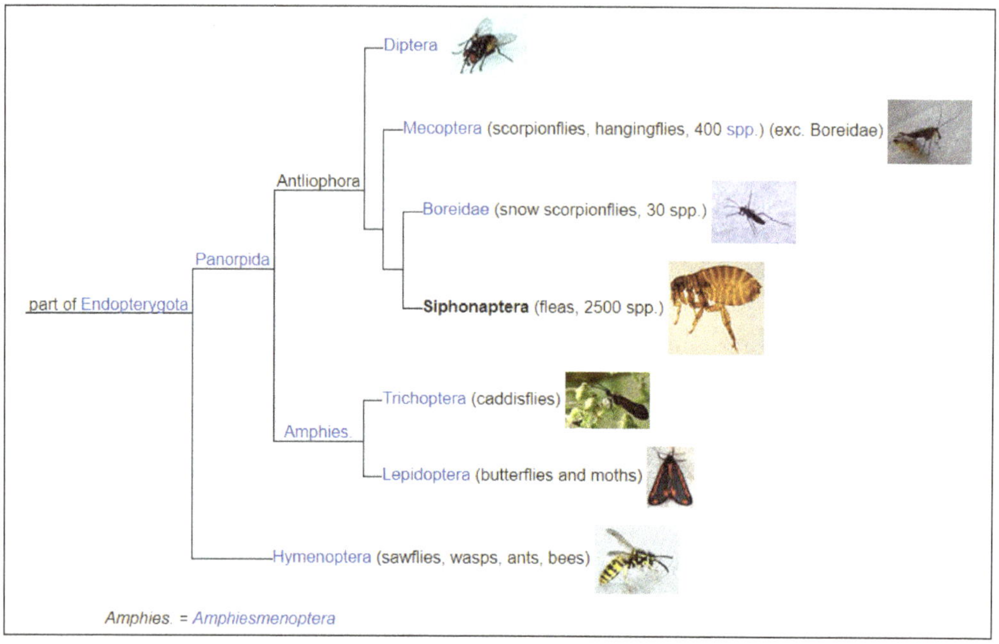

Fossils of wingless "pre-fleas" with siphonate (sucking) mouthparts from the middle Jurassic to early Cretaceous have been found in northeastern China. These belonged to three proposed extinct families, the Pseudopulicidae, the Saurophthiridae, and the Tarwiniidae. The last common ancestor of modern Siphonaptera separated from the Mecoptera during the early Cretaceous. Most flea families formed after the end of the Cretaceous (in the Paleogene and onwards). Fleas probably arose in the southern continental area of Gondwana, and migrated rapidly northwards from there. They most likely evolved with mammal hosts, only later moving to birds.

Siphonaptera is a relatively small order of insects: members of the order undergo complete metamorphosis and are secondarily wingless (their ancestors had wings which modern forms have lost). In 2005, Medvedev listed 2005 species in 242 genera, and despite subsequent descriptions of new species, bringing the total up to around 2500 species, this is the most complete database available. The order is divided into four infraorders and eighteen families. Some families are exclusive to a single host group; these include the Malacopsyllidae (armadillos), Ischnopsyllidae (bats) and Chimaeropsyllidae (elephant shrews).

Cenozoic flea in amber, c. 20 mya, is morphologically modern.

Many of the known species are little studied. Some 600 species (a quarter of the total) are known from a single record from a single host. Over 94% of species are associated with mammalian hosts, and only about 3% of species can be considered to be specific parasites of birds. The fleas on birds are thought to have originated from mammalian fleas; at least sixteen separate groups of fleas switched to avian hosts during the evolutionary history of the Siphonaptera. Occurrences of fleas on reptiles is accidental, and fleas have been known to feed on the hemolymph (bloodlike body fluid) of ticks.

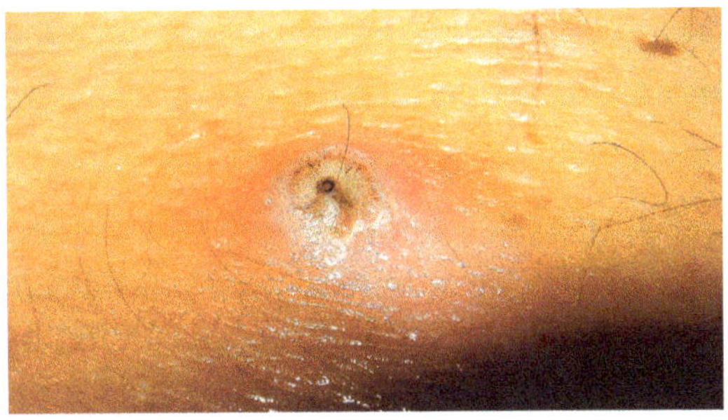

Jigger, Tunga penetrans, in human skin

Flea phylogeny was long neglected, the discovery of homologies with the parts of other insects being made difficult by their extreme specialization. Whiting and colleagues prepared a detailed molecular phylogeny in 2008, with the basic structure shown in the cladogram. The Tungidae, including the harmful chigoe flea or jigger, is sister to the rest of the Siphonaptera.

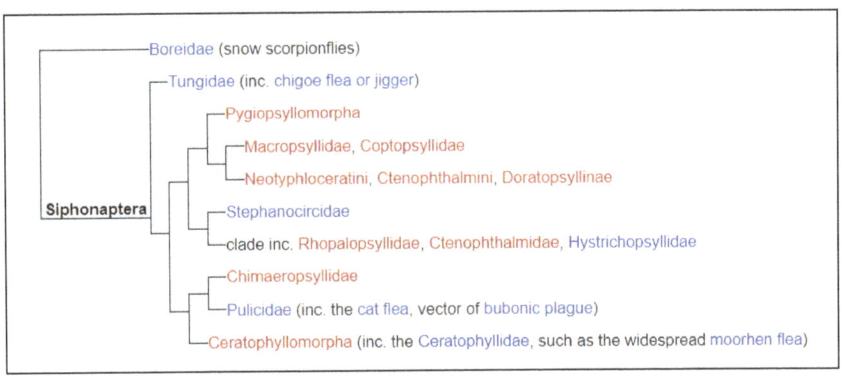

Relationship with Host

Fleas feed on a wide variety of warm-blooded vertebrates including humans, dogs, cats, rabbits, squirrels, ferrets, rats, mice and birds. Fleas normally specialise in one host species or group of species, but can often feed but not reproduce on other species. Ceratophyllus gallinae affects poultry as well as wild birds. As well as the degree of relatedness of a potential host to the flea's original host, it has been shown that avian fleas that exploit a range of hosts, only parasitise species with low immune responses. In general, host specificity decreases as the size of the host species decreases. Another factor is the opportunities available to the flea to change host species; this is smaller in colonially nesting birds, where the flea may never encounter another species, than it is in solitary nesting birds. A large, long-lived host provides a stable environment that favours host-specific parasites.

One theory of human hairlessness is that the loss of hair helped humans to reduce their burden of fleas and other ectoparasites.

Direct effects of Bites

In many species, fleas are principally a nuisance to their hosts, causing an itching sensation which in turn causes the host to try to remove the pest by biting, pecking or scratching. Fleas are not simply a source of annoyance, however. Flea bites cause a slightly raised, swollen itching spot to form; this has a single puncture point at the centre, like a mosquito bite. Besides this, the eczematous itchy skin disease flea allergy dermatitis is common in many host species, including dogs and cats. The bites often appear in clusters or lines of two bites, and can remain itchy and inflamed for up to several weeks afterwards. Fleas can lead to hair loss as a result of frequent scratching and biting by the animal, and can cause anemia in extreme cases.

As a Vector

Fleas are vectors for viral, bacterial and rickettsial diseases of humans and other animals, as well as of protozoan and helminth parasites. Bacterial diseases carried by fleas include murine or endemic typhus and bubonic plague. Fleas can transmit Rickettsia typhi, Rickettsia felis, Bartonella henselae, and the myxomatosis virus. They can carry Hymenolepiasis tapeworms and Trypanosome protozoans. The chigoe flea or jigger (Tunga penetrans) causes the disease tungiasis, a major public health problem around the world. Fleas that specialize as parasites on specific mammals may use other mammals as hosts; thus, humans may be bitten by cat and dog fleas.

Demodex

Demodex is a genus of tiny mites that live in or near hair follicles of mammals. Around 65 species of Demodex are known. Two species live on humans: Demodex folliculorum and Demodex brevis, both frequently referred to as eyelash mites. Different species of animals host different species of Demodex. Demodex canis lives on the domestic dog. Infestation with Demodex is common and usually does not cause any symptoms, although occasionally some skin diseases can be caused by the mites.

D. Folliculorum and D. Brevis

D. folliculorum and D. brevis are typically found on humans. D. folliculorum was first described in 1842 by Simon; D. brevis was identified as separate in 1963 by Akbulatova. D. folliculorum is found in hair follicles, while D. brevis lives in sebaceous glands connected to hair follicles. Both species are primarily found in the face - near the nose, the eyelashes, and eyebrows, but also occur elsewhere on the body.

The adult mites are only 0.3–0.4 mm (0.012–0.016 in) long, with D. brevis slightly shorter than D. folliculorum. Each has a semitransparent, elongated body that consists of two fused segments. Eight short, segmented legs are attached to the first body segment. The body is covered with scales for anchoring itself in the hair follicle, and the mite has pin-like mouthparts for eating skin cells and oils (sebum), which accumulate in the hair follicles. The mites can leave the hair follicles and slowly walk around on the skin, at a speed of 8–16 mm (0.31–0.63 in) per hour, especially at night, as they try to avoid light. The mites are transferred between hosts through contact with hair, eyebrows, and the sebaceous glands of the face.

Females of D. folliculorum are larger and rounder than males. Both male and female Demodex mites have a genital opening, and fertilization is internal. Mating takes place in the follicle opening, and eggs are laid inside the hair follicles or sebaceous glands. The six-legged larvae hatch after 3–4 days, and the larvae develop into adults in about 7 days. The total lifespan of a Demodex mite is several weeks.

Health Issues

Older people are much more likely to carry the mites; about a third of children and young adults, half of adults, and two-thirds of elderly people carried them. The lower rate in children may be because children produce less sebum. Recently, a study of n = 29 people in North Carolina, USA, found that all the adults (n = 19, over 18 years of age) carried mites, and that 70% of those under 18 years of age carried mites. This study (using a DNA detection method, more sensitive than traditional sampling and observation by microscope), along with several studies of cadavers, suggests that previous work might have underestimated the mites' prevalence. However, the small sample size and small geographical area involved prevent drawing broad conclusions from these data.

Research

Research about human infection by Demodex mites is ongoing:

- Evidence of a correlation between Demodex infection and acne vulgaris exists, suggesting it might play a role in promoting acne.
- Several preliminary studies suggest an association between mite infection and rosacea.

D. canis

D. canis

The natural host of D. canis is the domestic dog. Although it can temporarily infect humans, D. canis mites cannot survive on the human skin, and will die shortly after exposure and are considered not to be zoonotic. Naturally, the D. canis mite has a commensal relationship with the dog and under normal conditions does not produce any clinical signs or disease. The escalation of a commensal D. canis infestation into one requiring clinical attention usually involves complex immune factors. Under normal health conditions, the mite can live within the dermis of the dog without causing any harm to the animal. However, whenever an immunosuppressive condition is present and the dog's immune system (which normally ensures that the mite population cannot escalate to an infestation that can damage the dermis of the host) is compromised, it allows the mites to proliferate. As they continue to infest the host, clinical signs begin to become apparent and demodicosis/demodectic mange/red mange is diagnosed.

Since D. canis is a part of the natural fauna on a canine's skin, the mite is not considered to be contagious. All dogs receive an initial exposure from their mothers during nursing. The immune system of the healthy animal keeps the population of the mite in check, so subsequent exposure to dogs possessing clinical demodectic mange does not increase an animal's chance of developing demodicosis. Subsequent infestations after treatment can occur.

Demodex Folliculorum

Demodex folliculorum is a microscopic mite that can only survive on the skin of humans. Most people have D. folliculorum on their skin. Usually, the mites do not cause any harm, and are therefore considered an example of commensalism rather than parasitism. If D. folliculorum does cause disease, this is known as demodicosis.

Anatomy

D. folliculorum is adapted to live inside hair follicles, and therefore is thin and worm-like, with short legs. As an adult, D. folliculorum measures 0.3 to 0.4 mm (0.012 to 0.016 in) long. Adults have four pairs of legs, larvae and nymphs have only three pairs. D. folliculorum has a rudimentary gut, and no anus.

Reproduction and life Cycle

The entire life cycle of D. folliculorum takes 14–16 days. Adult mites copulate at the top of the hair follicle, near the skin surface. Eggs are deposited in the sebaceous gland inside the hair follicle. The heart-shaped egg is 0.1 mm (0.0039 in) long, and hatches into a six-legged larva. It takes seven days for the larva to develop into a mature adult, with two intervening nymph stages. The adult lives for 4–6 days.

Ecology

Demodex folliculorum prefers areas where sebum production is high, and is typically found in hair follicles on the human face, generally in greater numbers around the cheeks, nose, and forehead, but also elsewhere on the face, eyelids and ears. The mites may also be found on other parts of the body, such as the chest and buttocks.

Within the hair follicle, D. folliculorum is found above the sebaceous gland, positioned head downwards, with the end of abdomen often protruding from the hair follicle. Infested follicles usually contain 2–6 mites, but greater numbers can occur.

In one hour, D. folliculorum can travel 8 to 16 mm (0.31 to 0.63 in); the mites usually travel at night.

The mites are obligate commensals of humans, and can only live on the skin; they soon dry out and die if they leave the host. Higher numbers of D. folliculorum are found in the spring and summer than at other times of the year.

Relationship with Humans

D. folliculorum are not found on newborn babies, but are acquired shortly after birth, most likely due to maternal contact. Few mites are found on children under ten years of age, but nearly all elderly people are infested. The increasing population over time may be due to a small initial infestation gradually growing over time, or may be because levels of the mite's food, sebum, increase with age.

High numbers of D. folliculorum are associated with blepharitis and acne rosacea. The exact mechanism by which the mites cause disease is unknown; they may physically block the hair follicle, carry disease-causing bacteria, or, after death, their bodies may cause either a delayed hypersensitivity response, or an innate immune response. There is controversy over whether high numbers of D. folliculorum cause rosacea, or whether the skin environment caused by rosacea is more hospitable to mites than normal skin, allowing them to flourish. Populations of D. folliculorum are also increased in people with immunosuppression.

Tick

Ticks are arachnids, typically 3 to 5 mm long, part of the order Parasitiformes. Along with mites, they constitute the subclass Acari. Ticks are ectoparasites (external parasites), living by feeding on the blood of mammals, birds, and sometimes reptiles and amphibians. Ticks evolved by the Cretaceous period, the most common form of fossilisation being amber immersion. Ticks are widely distributed around the world, especially in warm, humid climates.

Almost all ticks belong to one of two major families, the Ixodidae or hard ticks, and the Argasidae or soft ticks. Adults have ovoid or pear-shaped bodies, which become engorged with blood when they feed, and eight legs. In addition to having a hard shield on their dorsal surfaces, hard ticks have a beak-like structure at the front containing the mouthparts, whereas soft ticks have their mouthparts on the underside of the body. Both families locate a potential host by odour or from changes in the environment.

Ticks have four stages to their lifecycle, namely egg, larva, nymph, and adult. Ixodid ticks have three hosts, taking at least a year to complete their lifecycle. Argasid ticks have up to seven nymphal stages (instars), each one requiring a blood meal. Because of their habit of ingesting blood, ticks are vectors of many diseases that affect humans and other animals.

Range and Habitat

Tick species are widely distributed around the world, but they tend to flourish more in countries with warm, humid climates, because they require a certain amount of moisture in the air to undergo metamorphosis, and because low temperatures inhibit their development from eggs to larvae. Ticks are also widely distributed among host taxa, which include marsupial and placental mammals, birds, reptiles (snakes, iguanas, and lizards), and amphibians. Ticks of domestic animals cause considerable harm to livestock by transmission of many species of pathogen, as well as causing anaemia and damaging wool and hides. Some of the most debilitating species occur in tropical countries. Tropical bont ticks affect most domestic animals and occur in Africa and the Caribbean. The spinose ear tick has a worldwide distribution, the young feeding inside the ears of cattle and wild animals.

In general, ticks are to be found wherever their host species occur. Migrating birds carry ticks with them on their journeys; a study of migratory birds passing through Egypt found more than half the bird species examined were carrying ticks. The species of tick often differed between the autumn and spring migrations, probably because of the seasonal periodicities of the different species.

For an ecosystem to support ticks, it must satisfy two requirements; the population density of host species in the area must be high enough, and humidity must be high enough for ticks to remain hydrated. Due to their role in transmitting Lyme disease, ixodid ticks, particularly the North American I. scapularis, have been studied using geographic information systems to develop predictive models for ideal tick habitats. According to these studies, certain features of a given microclimate – such as sandy soil, hardwood trees, rivers, and the presence of deer – were determined to be good predictors of dense tick populations.

A habitat preferred by ticks is the interface where a lawn meets the woods, or more generally, the ecotone, which is unmaintained transitional edge habitat between woodlands and open areas. Therefore, one tick management strategy is to remove leaf litter, brush, and weeds at the edge of the woods. Ticks like shady, moist leaf litter with an overstory of trees or at least shrubs, and they deposit their eggs into such places in the spring, so that the larvae can emerge in the fall and crawl into low-lying vegetation. The 3 m of boundary closest to the lawn's edge are a tick migration zone where 82% of tick nymphs in lawns are found.

Anatomy and Physiology

Ticks, like mites, are arthropods that have lost the segmentation of the abdomen that their ancestors had, with a subsequent fusion of the abdomen with the cephalothorax. The tagmata typical of other Chelicerata have been replaced by two new body sections, the anterior capitulum (or gnathosoma), which is retractable and contains the mouthparts, and the posterior idiosoma, which contains the legs, digestive tract, and reproductive organs. The capitulum is a feeding structure with mouthparts adapted for piercing skin and sucking blood; it is only the front of the head and contains neither the brain nor the eyes. Features of the capitulum include the basis capitulum, two palps, two cutting chelicerae, and hypostome. The basis capitulum supports the rest of the feeding structures. Palps have a sensory role and are composed of three sections. The hypostome is used for blood extraction and is a hollow, tube-like structure.

The ventral side of the idiosoma bears sclerites, and the gonopore is located between the fourth pair of legs. In the absence of segmentation, the positioning of the eyes, limbs, and gonopore on the idiosoma provide the only locational guidance.

Most ticks are inornate, meaning they are reddish or mahogany in color and lack markings. However, some species are ornate and have distinctive white patterns on the scutum.

Larval ticks hatch with six legs, acquiring the other two after a blood meal and molting into the nymph stage. In the nymphal and adult stages, ticks have eight legs, each of which has seven segments and is tipped with a pair of claws. The legs are sometimes ornamented and usually bear sensory or tactile hairs. In addition to being used for locomotion, the tarsus of leg I contains a unique sensory structure, Haller's organ, which can detect odors and chemicals emanating from the host, as well as sensing changes in temperature and air currents. Ticks can also use Haller's organs to perceive infrared light emanating from a host. When not being used for walking, the legs remain tightly folded against the body.

In nymphs and adults of the family Ixodidae, the capitulum is prominent and projects forwards from the body, a feature not present in the Argasidae. The eyes are close to the sides of the scutum, and the large spiracles are located just behind the coxae of the fourth pair of legs. The hard, protective scutellum, characteristic of this family, covers the whole dorsal surface in males, but is restricted to a small, shield-like structure behind the capitulum in females and nymphs. They differ, too, in their lifecycle; Ixodidae that attach to a host can bite painlessly and generally are unnoticed, and they remain in place until they engorge and are ready to change their skin; this process may take days or weeks. Some species drop off the host to moult in a safe place, whereas others remain on the same host and only drop off once they are ready to lay their eggs.

The body of the soft tick, family Argasidae, is pear-shaped or oval with a rounded anterior portion. The mouthparts cannot be seen from above, as they are on the ventral surface. The cuticle is leathery; often, a centrally positioned dorsal plate is seen, with ridges that project slightly above the surrounding surface, but with no decoration. A pattern of small, circular, depressed areas shows where muscles are attached to the interior of the integument. The eyes are on the sides of the body, the spiracles open between legs 3 and 4, and males and females only differ in the structure of the genital pore.

The Nuttalliellidae can be distinguished from both ixodid and argasid ticks by a combination of a projecting capitulum at the front and a soft, leathery skin. Other distinguish characteristics include the position of the stigmata, the lack of setae, the strongly corrugated integument, and the form of the fenestrated plates.

A hard-bodied tick of the family Ixodidae, the lone star tick.

Ticks are extremely tough, hardy, and resilient animals. They can survive in a near-vacuum for as long as a half hour. Their slow metabolism during their dormant periods enables them to go long periods between meals. During droughts, they can endure dehydration without feeding for as long as 18 weeks, although ticks with limited energy reserves may succumb to desiccation after 36 weeks. To keep from dehydrating, ticks hide in humid spots on the forest floor or absorb water from subsaturated air by secreting hygroscopic fluid produced by the salivary glands onto the external mouthparts and then reingesting the water-enriched fluid.

Ticks can withstand temperatures just above 0 °F (−18 °C) for more than two hours, and can survive temperatures in the 20–29 °F (−7 – −2 °C) range for at least two weeks. Ticks have even been found in Antarctica, where they feed on penguins.

A soft-bodied tick of the family Argasidae, beside eggs it has just laid.

Diet and Feeding

Ticks satisfy all of their nutritional requirements as ectoparasites, feeding on a diet of blood. They are obligate hematophages, needing blood to survive and move from one stage of life to another. Ticks can fast for long periods, but eventually die if unable to find a host. This behavior evolved approximately 120 million years ago through adaptation to blood-feeding. The behavior evolved independently in the separate tick families, with differing host-tick interactions driving the evolutionary change.

Some ticks attach quickly, while others wander around looking for thinner skin, such as is found on the ears of mammals. Depending on the species and the life stage, preparing to feed can take from 10 minutes to two hours. On locating a suitable feeding spot, the tick grasps the host's skin and cuts into the surface. They extract blood by cutting a hole in the host's epidermis, into which they insert their hypostome, and keep the blood from clotting by excreting an anticoagulant or platelet aggregation inhibitor.

Ticks find their hosts by detecting animals' breath and body odors, or by sensing body heat, moisture, and vibrations. They are incapable of flying or jumping, but many tick species, particularly Ixodidae, lie in wait in a position known as "questing". While questing, ticks cling to leaves and grasses by their third and fourth pairs of legs. They hold the first pair of legs outstretched, waiting to grasp and climb on to any passing host. Tick questing heights tend to be correlated with the size of the desired host; nymphs and small species tend to quest close to the ground, where they may encounter small mammalian or bird hosts; adults climb higher into the vegetation, where larger hosts may be encountered. Some species are hunters and lurk near places where hosts may rest. On receiving an olfactory or other stimulus, they crawl or run across the intervening surface.

Other ticks, mainly the Argasidae, are "nidicolous", finding hosts in their nests or burrows, and in caves in the case of bats. They use the same stimuli as non-nidicolous species to identify hosts, with body heat and odors often being the main factors. Many of them feed primarily on birds, though some Ornithodoros species, for example, feed on mammals. Both groups feed rapidly, typically biting painfully and drinking their fill within minutes. None of the species sticks to the host in the way that hard ticks do. Unlike the Ixodidae that have no fixed dwelling place except on the host, they live in sand or in crevices near animal dens or nests, or in human dwellings where they come out nightly to attack roosting birds, or emerge when they detect carbon dioxide in the breath of their hosts.

In the Ixodidae, the tick stays in place until it is completely engorged. Its weight may increase by 200 to 600 times as compared to its weight before it started feeding. To accommodate this large expansion, cell division takes place and its cuticle grows larger; the tick may remain attached for days or weeks, depending on species, life stage, and host. In the Argasidae, the tick's cuticle stretches to accommodate the fluid ingested, but does not grow new cells, with the weight of the tick increasing five- to 10-fold over the unfed state. The tick then drops off the host and typically remains in the nest or burrow until its host returns to provide its next meal.

Tick saliva contains about 1,500 to 3,000 proteins depending on the tick species. The proteins with anti-inflammatory properties, called evasins, help ticks to feed for 8–10 days without being noticed by the host animal, as they block the host's chemokines and prevent painful inflammation. Researchers are studying these evasins with the goal of developing drugs to neutralise the chemokines that cause inflammation in myocarditis, heart attack, and stroke.

Ecology

Mites and nematodes feed on ticks, which are also a minor nutritional resource for birds. More importantly, they carry diseases as the primary hosts of pathogens such as spirochaetes, and without their agency, the organisms could not reach their secondary hosts. The diseases caused may debilitate their victims, and ticks may thus be assisting in controlling animal populations and preventing overgrazing.

Certain infectious diseases of humans and other animals can be transmitted by ticks, with the species of tick involved tending to be those with a wide host range. Spread of disease in this way is enhanced by the extended time during which a tick remains attached, during which time the mobile host can be carried long distances, or in the case of bird hosts, across the sea. The infective agents can be present not only in the adult tick, but also in the eggs produced plentifully by the females. Many tick species have extended their ranges as a result of the movements of people, their pets, and livestock. With increasing participation in outdoor activities such as wilderness hikes, more people and their dogs may find themselves exposed to attack. Both ixodid and argasid ticks have four lifecycle stages: egg, larva, nymph, and adult.

Ixodidae

One-host Ticks

Eggs laid in the environment hatch into larvae, which immediately seek out a host in which to attach and feed. Fed larvae molt into unfed nymphs that remain on the host. After engorging on the host's blood, the nymphs molt into sexually mature adults that remain on the host in order to feed and mate. Once a female is both fed and ready to lay eggs, only then does she drop off the host in search of a suitable area to deposit her eggs. Ticks that follow this life cycle are called one-host ticks. The winter tick Dermacentor albipictus and the cattle tick Boophilus microplus are examples of one-host ticks.

Two-host Ticks

Newly hatched larvae attach to a host in order to obtain a blood meal. They remain on the host after developing into nymphs. After emerging from their shed larval skins, the nymphs reattach to the host and feed. Once engorged, they drop off the host and find a safe area in the natural environment in which to molt into adults. Both male and female adults seek out a host on which to attach, which may be the same body that served as host during their early development. Once attached, they feed and mate. Females ready to lay eggs drop from the host to oviposit in the environment. Ticks that complete their life cycle in this manner are called two-host ticks, like Hyalomma anatolicum excavatum.

Three-host Ticks

Most ixodid ticks require three hosts, and their lifecycles take at least a year to complete. Thousands of eggs are laid on the ground by an adult female tick. When the larvae emerge, they attach and feed primarily on small mammals and birds. After feeding, they detach from their hosts and molt to nymphs on the ground, which then attach and feed on larger hosts before dropping off yet again in order to molt into adults. Adults seek out a third host on which to feed and mate. Female adults engorge on blood and prepare to drop off to lay her eggs on the ground, while males feed very little and remain on the host in order to continue mating with other females.

Argasidae

Argasid ticks, unlike ixodid ticks, may go through up to seven nymphal stages (instars), requiring a meal of blood each time. Their lifecycles range from months to years. The adult female argasid tick can lay a few hundred to over a thousand eggs over the course of her lifetime. Larvae feed very quickly and detach to molt into nymphs. Both male and female adults feed on blood, and they mate off the host. During feeding, any excess fluid is excreted by the coxal glands, a process that is unique to argasid ticks.

Parasitic Bugs

Bat Bug

Bat bugs are blood-sucking insect parasites that feed primarily on the blood of bats. The name has been applied to members of the family Cimicidae (e.g. Cimex lectularius, Afrocimex constrictus) and also to members of the family Polyctenidae. Bat bugs are closely related to bed bugs, and are so similar in appearance that they are often mistaken for bed bugs. Microscopic examination is needed to distinguish them. Bat bugs will also bite humans if given the opportunity.

Bat bug species include:

- African bat bug (Afrocimex constrictus),
- Eastern bat bug (Cimex adjunct).

Anatomy

A key physiological distinction between the common bedbug and the bat bug is the fringe hairs on the pronotum (the upper covering of the thorax), which are at least as long as the width of the bat bug's eye, but shorter in the bedbug.

Infestations

Bat bugs are moderately common in the midwest US and have been recorded in Scotland, and are found in houses and buildings that harbor bats. Infestations in human dwellings are usually introduced by bats carrying the bugs on their skin. Bat bugs usually remain in close proximity to the roosting locations of bats (attics, chimneys, etc.) but explore the rest of the building if the bats leave or are eliminated. In some cases, they move into harborages that are more typical of bedbugs, such as mattresses and bed frames.

Life Cycle

Development from egg to adult ranges from 2 weeks in ideal conditions (warm temperature and abundant food supply) to more than 15 weeks, averaging about 1.5 months. An adult may survive more than one year without feeding. As with the common bedbug, a nymph requires a blood meal to molt, and an adult female requires a blood meal to lay eggs.

Feeding Habits

Bat bugs feed on blood from bats, but when they wander away from the bat roost area, they will feed on other warm-blooded animals, including people. This feeding is an annoyance but is not dangerous. Bat bugs have not been found to transmit any diseases.

Bed Bug

Bedbugs are small, wingless insects that feed exclusively on the blood of warm-blooded animals. Humans are the preferred hosts for the two main species.

There are two species of bedbugs that are known to feed on human blood. They are known scientifically as Cimex lectularius and Cimex hemipterus. They have been found in the tombs of ancient Egyptians from 3,500 years ago.

Over millions of years, bedbugs have evolved as nest parasites, inhabiting the nests of birds and the roosts of bats. Some of them have learned to adapt to the human environment.

Newborn bedbugs, called hatchlings or nymphs, are tiny but visible and about the size of a poppy seed. Adults grow to about 0.25 inches long with an oval and flattened shape when they are not feeding. After feeding, they can double in size. Nymphs, eggs, and adults are visible to the naked eye.

They are called bedbugs because of their preferred habitat in human homes: Sofas, bed mattresses, clothing, and other soft furnishings. They also prefer the dark.

Bedbugs are seen as a growing problem within all types of dwellings, including private homes, dormitories, cruise ships, army barracks, and shelters.

When seen close up, their color may range from a white, light tan to a deep brown or burnt orange color. When they have fed, a dark red or black blob may be observed within their body. They seek shelter in dark cracks and crevices when disturbed.

Signs

The most obvious sign of bedbugs in the home is that people complain of bites that occurred while they were asleep. If this happens, examine the bedrooms for bedbugs and signs of bedbug activity.

Look carefully in bed linen and the seams and tufts of mattresses and box springs for bugs or eggs. The eggs will look like tiny, pale poppy seeds. Signs of bedbug activity may occur beneath loose areas of wallpaper near beds, in the corner of desks and dressers, in the laundry, and in drawers.

Keep an eye out for dark brown or rust-colored bedbug droppings that stain material and mattresses. Bedbug excrement is a liquid that looks either light brown or black, and it usually either beads up or is absorbed by the material around it. A large population of bedbugs may produce a coriander-like odor.

A bedbug may cause a coriander-like smell.

Treatment

Treatment focuses on relieving symptoms and includes:

- Topical creams, such as cortisone, to relieve itching.
- An Oral antibiotic, if infection occurs because of skin irritation around the bite.
- Corticosteroids, if a person has a severe allergic reaction.
- Antihistamines, to help relieve allergic reactions.

Most bites heal within 1 and 2 weeks of occurrence.

Some of these treatments are available for purchase online, including cortisone and antihistamines.

Removal

Since bedbugs can hide in a wide range of places in the home, they are not easy to remove. It is advisable to bring in a pest control professional.

Removing excess clutter from the house, giving the bedbugs fewer places to hide, makes inspection and removal less difficult.

Some pest control companies request that furniture is pulled away from walls and mattresses and box springs stood on edge before they enter the home. Other companies prefer everything to be left where it is so that they can perform a check before moving the furniture themselves.

Scientists at Ohio State University have determined that combining the chemical signals of bedbugs with a common insect-control agent can make it an effective treatment for killing the bugs.

Symptoms

Most bedbugs feed on their hosts while they sleep. They draw blood in a painless way. While feeding, they inject a small amount of saliva into the host's skin. If they feed on one particular person for several weeks, the individual may become more sensitive to their saliva and the chemicals that it contains. The host might eventually develop an allergic response.

Bedbugs, like fleas, tend to bite in rows. There are likely to be two or three bites in each row. This is probably because the bedbug is interrupted while feeding, and then comes back about half an inch further down for its next bite. Bites can take up to 14 days to become visible but often appear within several days. Bedbug bites are larger than fleabites and do not usually have a red dot at the center. The bites tend to be raised and red.

They can be scattered or occur in clusters of three over the paths of blood vessels, known as the "breakfast, lunch, and dinner sign."

Most people who are bitten show no symptoms at all and often do not know it happened. This makes it more difficult to prevent or identify potential infestations. Some individuals, however, may become ill and nauseous. It is possible to get scars and skin infections from scratching the bites. Very rarely, people might have an anaphylactic reaction to bedbug bites. It is possible but rare to have an asthmatic reaction to bedbugs.

Causes

Bedbugs are adaptable, and there are many ways in which a bedbug infestation can occur. They may get into a new home as stowaways when luggage, furniture, and bedding is moved in. People should be careful when purchasing second-hand furniture and should never purchase used mattresses. A careful visual inspection should allow a person to detect bedbugs or their droppings.

Even vacant and seemingly clean homes may have bedbugs in them. They can survive for over two months without any food. It is also believed they can move from apartment to apartment through hollows and holes in the walls and the tubes through which wires and pipes run.

Prevention

Bedbug infestations can be difficult to prevent. It is possible to encase both the mattress and box spring in a protective cover, as some people do for allergy relief. Some pest control firms sell them, as well as a number of retail outlets.

Once encased, any bedbug trapped inside and prevented from feeding will eventually die. Some people keep their new beds encased, as it prevents the bugs from getting into the crevices in the mattress and makes it easier to keep the surface clean and bug-free.

When traveling, avoid putting luggage on the bed to reduce the risk of bringing bedbugs home in a suitcase. It is also worth vacuuming any luggage after you return home and making sure you get rid of the contents of the vacuum in a tightly sealed bag. Dispose of this bag in a trash-can outside.

Separate travel clothes from laundry and immediately wash them in hot water. Although they are not known to carry diseases, bedbugs can affect an individual's quality of life, causing distress, discomfort, embarrassment, and broken sleep.

Cimicidae

The Cimicidae are a family of small parasitic insects that feed exclusively on the blood of warm-blooded animals. They are called cimicids or, loosely, bed bugs (or bedbugs or bed-bugs), though the latter term properly refers to the most famous species of the family, Cimex lectularius, the common bed bug. Around 90 species are placed in the family Cimicidae.

Biology

All cimicids are small, oval-shaped, and flat in appearance, although their bodies bulge after feeding. They do not fly, but do have small, nonfunctional wing pads. They also have beak-like mouth parts with which they pierce the skin and suck the blood of their hosts.

Cimicids practice traumatic insemination. Although the female has a normal genital tract for laying eggs, the male never uses it (except in the species Primicimex cavernis), instead piercing the female's abdominal wall; the sperm then migrate through the female's paragenital system.

Feeding is required for egg production in females and possibly for sperm production in males. Egg-laying behavior varies among species. C. lectularius stops laying fertile eggs about 35 to 50 days after the last insemination. The swallow bug, Oeciacus vicarius, hibernates after mating in autumn and begins laying in spring, to coincide with the return of the host.

The five nymphal instars (stages) must each take a blood meal to develop to the next stage. Adults have been reported to live 3–12 months in an untreated household situation. A given undisturbed bug may take 3–15 minutes to take a full meal depending on its life stage. They can survive long periods of time without feeding, reappearing from their hiding places when hosts again become available. They have special scent glands and often emit a characteristic odor that can usually be detected in heavily infested sites. Dark fecal spots on bedding or bedclothes, pearly white eggs (1 mm long), and papery yellowish cast "skins" are other signs of infestation often found in cracks or crevices. Cimicidae are easily repelled by excessive heat.

Hosts

Cimicids are relatively specialized in their choice of hosts, compared to other bloodsucking insects. Most cimicids have a preferred host, but accept some others when presented with the choice, such as C. lectularius and C. hemipterus, which are most often found among humans, but can also survive by feeding on birds, bats, rabbits, and mice. Some subfamilies are restricted to certain types of bats, while one species, P. cavernis, appears to accept only one species of host.

Host switching is dependent on several factors, including overlap in host detection cues and ability to digest different kinds of blood. For example, the red blood cells of chickens are about 3 to 5 μm longer in diameter than those of humans, making human blood more suitable for the narrow food canal of C. lectularius. C. hemipterus may be able to vary the size of its food canal, allowing it greater flexibility in its choice of hosts. Preference for a host species can vary between populations of a given species; the causes for this are unclear.

Behavior

Cimicids are attracted to hosts by a variety of cues, including heat (even a temperature difference of 1 °C) and kairomones. Host cues (at least in some species, including C. lectularius and Stricticimex antennatus) change from attractants to repellants after a cimicid has fed, causing it to move out of a danger zone after feeding.

Most cimicids feed once every 3 to 7 days in natural conditions. C. lectularius normally feeds once every 7 days and Ornithocoris toledoi every 8 days, though C. hemipterus has been observed feeding every day for several days (in hot climates). Excessively hot or cold temperatures disrupt normal behavior.

Transmission of Pathogens

Although viruses and other pathogens can be transmitted to cimicids, they rarely transmit them to their hosts. O. vicarius is a vector of several arboviruses, but is not killed by these viruses. Trypanosoma cruzi, the trypanosome that causes Chagas disease, is rarely transmitted from cimicids to bats, but it has not been observed replicating after such transmission. The viruses HIV and hepatitis B can persist in C. lectularius for two weeks, but with no viral replication. The possibility of these and most other viruses being transmitted from C. lectularius to humans is considered extremely remote.

References

- Prundeanu Croitoru, Anca G.; Chen, Helen M.; Ramos-e-Silva, Marcia; Busam, Klaus J. (2010). "Chapter 3. Infectious diseases of the skin. Demodicosis". In Busam, Klaus J. (ed.). Dermatopathology. Saunders. p. 171. ISBN 978-0443066542
- Arthropod-intoduction, contents , para-site: parasite.org.au, Retrieved 14 June, 2019

- Taylor, Sean D.; Cruz, Katharina Dittmar de la; Porter, Megan L.; Whiting, Michael F. (May 2005). "Characterization of the Long-Wavelength Opsin from Mecoptera and Siphonaptera: Does a Flea See?". Molecular Biology and Evolution. 22 (5): 1165–1174. Doi:10.1093/molbev/msi110. ISSN 0737-4038. PMID 15703237
- Krasnov, Boris R. (2008). Functional and Evolutionary Ecology of Fleas: A Model for Ecological Parasitology. Cambridge University Press. pp. 44–54. ISBN 978-1-139-47266-1
- Yeates, David K.; Wiegmann, Brian. "Endopterygota Insects with complete metamorphosis". Tree of Life. Retrieved 24 May 2016

6

Intracellular Parasites

The microparasites that can grow and reproduce inside the cells of a host are known as intracellular parasites. Cystoisospora belli, coccidian and toxoplasma are some examples of intracellular parasites. This chapter closely examines these key types of intracellular parasites to provide an extensive understanding of the subject.

If the parasitic organisms penetrate the host cells, they are said to be intracellular parasites, and can be further divided into two types - obligate and facultative.

Facultative parasite is an organism that does not rely on the host for completing its life cycle, but displays parasitism to derive nutrition from the host. In other words, it grows and reproduces on its own, but depends on the host cell as a nutrient and energy source. In contrast, an obligate parasite not only depends on the host for nutrition, but also for the purpose of reproduction. These organisms infect the host cell and utilize the internal-cellular machinery of the host to replicate themselves. After the parasite has reproduced itself multiple times, the progeny are released by rupturing the host cell's membrane. This leads to the lysis (death) of the host cell.

Characteristics of Intracellular Parasite

Species

This form of parasitism is exhibited by a wide range of organisms that include viruses, bacteria, fungi, and mold. They infect other organisms in order to complete their life cycle. This is so, because the progression of their life stages is highly specific and dependent on the host organisms. Some complete their life cycles within a single host and are said to have direct life cycles (e.g., Necator americanus), whereas few others need at least 2 or more hosts in a particular order, and they are said to possess indirect or complex life cycles (e.g., Plasmodium falciparum).

Nutrition

Since the organisms depend on the host cell for nutrition, they must keep it alive and viable for as long as possible till their own life cycle has been completed. This is seen in the case of lytic viral infection of a host cell. The virus infects the cell and takes over the cellular functioning, but it does not harm the cell till it has reproduced and replicated repeatedly. This goes on till the sheer number of the progeny causes the cell membrane to rupture (cell explodes) and release the offspring.

Invasion of Host Cell

A variety of mechanisms are employed by the obligate parasites to invade the host cell and then evade its immune response. Since the parasites are usually smaller in size, they rely on either passive or active uptake by the host cell. Viruses only insert their genetic material in the host cell, where it gets incorporated into the host's genome. Other parasites bear markers on their surface which are similar to the ones on the surface of host cells. Since they evade the immune system, the presence and proliferation of these parasites gives rise to a multitude of health disorders within the host.

Change in Host Behavior

The infection by an obligate parasite that follows a complex life cycle modifies the cellular chemistry of the host. This in turn allows the parasite to manipulate the behavior of the host in order to facilitate its transfer to the next host required to complete its life cycle. This is observed in the case of Toxoplasma gondii. It requires two hosts for the completion of its life cycle. Any warm-bodied animal can be used as the initial host during its asexual phase. Mostly rats are carriers of this parasite since they live in sewers and ingest fecal matter. The parasite induces the production of a chemical within the rat that causes the rat to be attracted to (instead of being repelled by) the odor of cat urine. This attraction causes it to seek out places populated with cats, and cat being the natural predator of rat, preys on the infected rat while also consuming the parasite. Here, in the cat's system, the asexual parasite completes the sexual part of its life cycle.

Examples:

- Viruses
  - All viruses are obligate intracellular parasites as they cannot reproduce outside the host cell, e.g., bacteriophages, AIDS virus (HIV), human papilloma virus (HPV), tobacco mosaic virus (TMV), etc., at all.
- Bacteria
  - Sarcobium lyticum infects small amoebae.
  - Chlamydia trachomatis causes a sexually-transmitted disease of the same name as its genus, in humans. It also causes nongonococcal urethritis, pelvic inflammatory disease, and trachoma.
  - Rickettsia prowazekii infection causes typhus fever in humans.
  - Inhalation of Mycobacterium tuberculosis causes tuberculosis, a disease that affects the lungs.
- Fungi
  - 14 genera of microsporidia parasites infect a variety of fish. For example, Glugea infects flounder, Loma infects Atlantic cod, salmon, and rainbow trout, Nosemoides infects cichlids, etc.
  - Pneumocystis jirovecii infects humans and causes pneumonia.

- Protozoa
    - Plasmodium falciparum causes malaria in humans.
    - Leishmania sps cause leishmaniasis in humans.

Despite having recognized the type of parasitism exhibited by these organisms, the exact mechanism employed by them to invade the host cell and evade elimination via the immune system, is yet to be fully deciphered. The knowledge of these mechanisms would prove immensely beneficial in understanding the regulation of cellular mechanisms and subsequently applying them in the fields of personalized medicine, biomimicry, etc.

Coccidia

Coccidia (Coccidiasina) are a subclass of microscopic, spore-forming, single-celled obligate intracellular parasites belonging to the apicomplexan class Conoidasida. As obligate intracellular parasites, they must live and reproduce within an animal cell. Coccidian parasites infect the intestinal tracts of animals, and are the largest group of apicomplexan protozoa.

Infection with these parasites is known as coccidiosis. Coccidia can infect all mammals, some birds, some fish, some reptiles, and some amphibians. Most species of coccidia are species-specific in their host. An exception is Toxoplasma gondii, which can infect all mammals, although it can only undergo sexual reproduction in cats. Depending on the species of coccidia, infection can cause fever, vomiting, diarrhea, muscle pain, and nervous system effects and changes to behavior, and may lead to death. Healthy adults may recover without medication—but those who are immunocompromised or young almost certainly require medication to prevent death. Humans generally become infected by eating under-cooked meat, but can contract infection with T. gondii by poor hygiene when handling cat waste.

Taxonomy

The class is divided into four orders, which are distinguished by the presence or absence of various asexual and sexual stages:

- Order Agamococcidiorida Levine-1979.
- Order Eucoccidiorida Léger and Duboscq-1910.
- Order Ixorheorida Levine-1984.
- Order Protococcidiorida Kheisin-1956.

The order Eucoccidiorida is divided into two suborders. These two groups differ in their sexual development: syzygy for Adeleorina and independent gametes for Eimeriorina.

The first suborder, Adeleorina, comprises coccidia of invertebrates and the coccidia that alternate between blood-sucking invertebrates and various vertebrates; this group includes Haemogregarina and Hepatozoon. There are seven families in this suborder.

The second suborder, Eimeriorina, comprises coccidia of a variety of coccidia many of form cysts. A number of genera, including Toxoplasma and Sarcocystis, infect vertebrates.

Coccidiosis

Transmission

Infected animals spread spores called oocysts in their stool. The oocysts mature, called sporulation. When another animal passes over the location where the feces were deposited, they may pick up the spores, which they then ingest when grooming themselves. The spores may also be ingested by mice; when another animal eats the mouse it becomes infected.

Some species of coccidia are transmissible to humans, including toxoplasma and cryptosporidium.

Infection

Inside the host, the sporulated oocyst opens, and eight sporozites are released. Each one finds a home in an intestinal cell and starts the process of reproduction. These offspring are called merozoites. When the cell is stuffed full of merozoites, it bursts open, and each merozoite finds its own intestinal cell to continue the cycle.

Symptoms of Infection

As the infection continues, millions of intestinal cells may become infected. As they break open, they produce a bloody, watery diarrhea. This can cause dehydration, and can lead to death in young or small pets.

Diagnosis and Treatment

Coccidiosis can be diagnosed by finding oocysts in fecal smears. In early stages of the disease, there may be very few oocysts being shed, and a negative test does not rule out the disease.

Coccidiosis is most commonly treated through the administration of coccidiostats, a group of medications that stop coccidia from reproducing. In dogs and cats, the most commonly administered coccidiostat is sulfa-based antibiotics. Once reproduction stops, the animal can usually recover on its own, a process that can take a few weeks, depending on the severity of the infection and the strength of the animal's immune system.

Toxoplasma Gondii

Toxoplasma gondii is an intestinal coccidium that parasitizes members of the cat family as definitive hosts and has a wide range of intermediate hosts. Infection is common in many warm-blooded animals, including humans. In most cases infection is asymptomatic, but devastating disease can occur. Toxoplasmosis is an infection caused by Toxoplasma gondii.

It can be found in cat feces and undercooked meat, especially venison, lamb, and pork. It can also be transmitted through contaminated water. Toxoplasmosis can be deadly or cause serious birth defects for a fetus if the mother becomes infected. This is why doctors recommend against pregnant woman scooping or cleaning cat litter boxes.

Most people who have toxoplasmosis never have any symptoms at all. According to the Centers for Disease Control and Prevention (CDC) over 60 million people in the United States are infected with the parasite. The people who are most at risk for serious infections are those with compromised immune systems and infants born to mothers with active infection during their pregnancy.

Symptoms of Toxoplasmosis

Most people who've been infected with the parasite that causes toxoplasmosis show no signs or symptoms.

People who develop symptoms may experience:

- A fever,
- Swollen lymph nodes, especially in the neck,
- A headache,
- Muscle aches and pains,
- Sore throat.

These symptoms can last for a month or more and usually resolve on their own.

Toxoplasmosis is especially serious for people who have weakened immune systems. For these people, they're at risk of developing:

- Brain inflammation, causing headaches, seizures, confusion and coma.
- A lung infection, causing cough, fever, and shortness of breath.
- An eye infection, causing blurry vision and eye pain.

When a fetus is infected, the symptoms may be mild or quite serious. Toxoplasmosis in an unborn baby can be life-threatening for the baby soon after birth. Most newborns with congenital toxoplasmosis may appear normal at birth but can develop signs and symptoms as they age. It's particularly important to check for involvement in their brain and eyes.

Causes of Toxoplasmosis

T. gondii is theparasite that causes toxoplasmosis. You can catch it from contaminated meat that's raw or not thoroughly cooked. You can also get toxoplasmosis by drinking contaminated water. In rare cases, toxoplasmosis may be transmitted through a blood transfusion or a transplanted organ.

The parasite can also exist in feces. This means it can be found on some unwashed produce that has been contaminated with manure. Wash your produce thoroughly to prevent toxoplasmosis.

In the United States, the parasite is found in cat feces. Although T. gondii is found in nearly all warm-blooded animals, cats are the only known hosts. This means that the parasite's eggs only reproduce sexually in cats. The eggs exit the feline's body through excretion. Cats don't usually show symptoms of toxoplasmosis even though they're hosts.

People become infected with toxoplasmosis only if they ingest the parasite. This could happen when being exposed to contaminated cat feces. This is most likely when cleaning out a litter box without washing your hands afterward.

Pregnant women have an increased risk of passing toxoplasmosis to their unborn child in this manner. For this reason, you should ask someone else to take care of the cat litter box during your pregnancy. If you absolutely must clean out the box yourself, protect yourself with gloves and change the cat litter box daily. The parasite isn't infectious until one to five days after it's shed.

It's very rare for humans to get toxoplasmosis from cats. Generally speaking, house cats that aren't allowed outside don't carry T. gondii. Wild cats or cats that live outside and hunt are more likely to be hosts of T. gondii.

In the United States, the most common way to get infected with the toxoplasmosis parasite is by eating raw meat or unwashed fruits and vegetables.

Toxoplasmosis Diagnoses

Your doctor will typically perform a blood test to check for antibodies to this parasite. An antibody is a type of protein that your immune system produces when it's threatened by harmful substances. Antibodies detect foreign substances by their surface markers, called antigens. Antigens include:

- Viruses,
- Bacteria,
- Parasites,
- Fungi.

Once an antibody has developed against a particular antigen, it will remain in your bloodstream to protect against future infections with that particular foreign substance.

If you've ever been exposed to T. gondii, antibodies will be present in your blood. This means you will test positive for the antibodies. If your tests come back positive, then you've been infected with this disease at some point in your life. A positive result doesn't necessarily mean that you currently have an active infection.

If your tests come back positive for antibodies, your doctor might do further testing to help figure out exactly when you were infected.

If you're pregnant and have an active infection, your doctor may test your amniotic fluid and the fetus' blood. An ultrasound can also help determine whether the fetus has been infected.

If your fetus is diagnosed with toxoplasmosis, you'll probably be referred to a specialist. Genetic counseling will also be suggested. The option of ending the pregnancy, depending on the gestational age of the baby, may be offered as a possibility. If you continue the pregnancy, your doctor will likely prescribe antibiotics to help reduce your baby's risk of symptoms.

Complications are Associated with Toxoplasmosis

The reason that pregnant woman should take special precautions to avoid toxoplasmosis is that it can be very serious, even fatal, for a baby infected in the uterus. For those who survive, toxoplasmosis can have lasting consequences on the:

- Brain,
- Eyes,
- Heart,
- Lungs.

They may also have mental and physical developmental delays and recurring seizures.

In general, babies who are infected early during the pregnancy suffer from more severe issues than those infected later in the pregnancy. Babies born with toxoplasmosis may have a higher risk of hearing and vision losses. Some children may be affected with learning disabilities.

Toxoplasmosis Treatment

Your doctor may recommend not treating your toxoplasmosis if it isn't causing any symptoms. Most healthy people who develop an infection don't have any symptoms or develop mild symptoms that are self-limited.

If the disease is severe, is persistent, involves the eyes, or involves the internal organs, your doctor will typically prescribe pyrimethamine (Daraprim) and sulfadiazine. Pyrimethamine is also used to treat malaria. Sulfadiazine is an antibiotic.

If you have HIV or AIDS, you may need to continue these medications for life. Pyrimethamine decreases your levels of folic acid, which is a type of B vitamin. Your doctor may also ask you to take additional vitamin B while taking the drug.

Treatment during Pregnancy

Treatment during pregnancy is somewhat different. Your course of treatment will depend on whether your unborn child is infected and the severity of the infection. Your doctor will talk with you about the best course for your particular case. Most likely, you'll be prescribed an antibiotic according to how far along you are in your pregnancy to reduce the likelihood of transmission to the fetus. An antibiotic called spiramycin is generally recommended in the first and early second trimester. A combination of pyrimethamine/sulfadiazine and leucovorin is generally used during the late second and third trimesters.

If your unborn baby has toxoplasmosis, pyrimethamine and sulfadiazine may be considered as a treatment. However, both drugs have significant side effects on women and the fetus and are only used as a last resort. The potential side effects include suppression of the bone marrow that helps produce blood cells and liver toxicity.

Outlook for People with Toxoplasmosis

The outlook for people with this condition depends on several factors. Pregnant women who develop this condition will need to work with their doctor to come up with a treatment plan that's right for them. Babies born with toxoplasmosis may receive treatments for up to a year.

People with AIDS and children with compromised immune systems may need to be hospitalized for treatment to prevent complications.

If you aren't pregnant and you don't have any underlying health conditions you should recover in several weeks. Your doctor may not prescribe any treatments if your symptoms are mild and you're otherwise healthy.

Toxoplasmosis Prevention

You can prevent toxoplasmosis by:

- Washing all fresh produce before you eat it.
- Making sure all meat is properly cooked.
- Washing all utensils that are used to handle raw meat.
- Washing your hands after cleaning or scooping cat litter.

Pregnant women should have someone else clean the cat litter box during their pregnancy.

Cystoisospora Belli

Cystoisospora belli, previously known as Isospora belli, is a parasite that causes an intestinal disease known as cystoisosporiasis. This protozoan parasite is opportunistic in immune suppressed human hosts. It primarily exists in the epithelial cells of the small intestine, and develops in the cell cytoplasm. The distribution of this coccidian parasite is cosmopolitan, but is mainly found in tropical and subtropical areas of the world such as the Caribbean, Central and S. America, India, Africa, and S.E. Asia. In the U.S., it is usually associated with HIV infection and institutional living.

Morphology

A fully mature (sporulated) oocyst of genus Isospora is a spindle-shaped body that has two sporocysts that contain four sporozoites each. The oocysts of Cystoisospora belli are long and oval shaped. They measure between 20 and 33 micrometers in length and between 10 and 19 micrometers wide.

Life Cycle

- An oocyst with one sporoblast is released in stool of infected person.
- After the oocyst has been released, the sporoblast matures further and divides into two.

- After the sporoblasts divide they create a cyst wall and become sporocysts.
- The sporocysts each divide twice, resulting in four sporozoites.
- Transmission occurs when these mature oocysts are ingested.
- The sporocysts excyst in the small intestine where sporozoites are released.
- The sporozoites then invade epithelial cells and schizogony is initiated.
- When the schizonts rupture, mereozoites are released and continue to invade more epithelial cells.
- Trophozoites develop into schizonts, containing many mereozoites.
- After about one week, development of male and female gametocytes begin in the mereozoites.
- Fertilization results in the development of oocysts, which are released in the stool.

The sporulation time of this parasite's egg is usually 1–4 days, and the entire life cycle takes about 9–10 days. The infective stage found in stool is the mature oocyst. The mature oocyst for Cystoisospora belli can remain infective in the environment for months.

Symptoms of Cystoisospora Belli

Immune competent individuals are usually asymptomatic to this parasite's infection. But clinical symptoms such as mild diarrhea, abdominal discomfort, and low grade fever for approximately one week has been observed in some individuals.

Immunocompromised people are more severely affected by Cystoisospora belli and can experience extreme diarrhea that can lead to weakness, anorexia, and weight loss. Other symptoms of cystoisosporiasis include abdominal pain, cramps, loss of appetite, nausea, vomiting, and fever, that can last from weeks to months.

Diagnosis and Treatment

Cystoisospora belli is diagnosed by identification of the oocyst through examining a stool sample under a microscope. The diagnostic stage is the immature oocyst that contains a spherical mass of protoplasm.

In other words, the oocyst that is diagnosed in the stool sample is unsporulated, and contains only one sporoblast. For stool diagnosis, direct smear, concentration smear, microscopic wet mount, or iodine stains of fecal smears are adequate. But for easy screening, acid-fast stains is recommended. If stool test is negative, and biopsies of the small intestine is performed, different stages of schizogony and sporogony should exist in the epithelial cells, but the alteration of the villi is not necessarily present. Eosinophilia may also be seen unlike in the case of other protozoal infections.

This infection is easily treated with antibiotics. The most common antibiotic that is prescribed is co-trimoxazole (trimethoprimsulfamethoxazole), more commonly known as Bactrim, Septra, or Cotrim.

In AIDS patients, treatments can result in the disappearance of the symptoms, but recurrence of symptoms is common. In order prevent the recurrence, medication is continued in AIDS patients and other immunosuppressed patients.

Transmission and Prevention

Cystoisospora belli does not require an intermediate host and currently is only known to transmit from person to person. The method of transmission is ingesting food or water that has been contaminated with feces from someone who is infected. Washing your hands with soap and warm water after using the toilet, changing diapers, and before handling food is vital. Also, educating children the importance of hand-washing and good hygiene practice is important. Because HIV-AIDS patients will have higher risk of symptomatic intestinal parasitic infections, and pathogenic burden can increase disease progression and contributes to early death, routine screening of parasites especially in patients with lower CD4 count should be emphasized.

References

- Obligate-intracellular-parasite-definition-characteristics-examples: biologywise.com, Retrieved 15 July, 2019
- "Biodiversity explorer: Apicomplexa (apicomplexans, sporozoans)". Iziko Museums of Cape Town. Archived from the original on 2006-09-23. Retrieved 2006-10-13
- Gupta, K., Bala, M., Deb, M., Muralidhar, S., & Sharma, D. K. (2013). Prevalence of intestinal parasitic infections in HIV-infected individuals and their relationship with immune status. Indian Journal of Medical Microbiology, 31(2), 161-165. Doi:10.4103/0255-0857.115247
- Toxoplasmosis-causes, health: healthline.com, Retrieved 16 August, 2019
- Auwaerter, Paul. "Cystoisospora belli". Johns Hopkins Guides. Johns Hopkins Medicine. Retrieved 20 April 2015

7
Parasites in Plants

Parasitic plants are the plants that obtain their nutritional requirement from other living plants. Dodder, nuytsia, orobanche minor, viscum album and orobanche unifloraare are some of the common examples of parasitic plants. This chapter has been carefully written to provide an easy understanding of these parasitic plants.

Dodder

Dodder is a group of ectoparasitic plants with about 150 species in a single genus, Cuscuta, in the morning glory family (Convolvulaceae) or Cuscutaceae, depending on the classifi cation system used. These pale green or yellow, bright orange or red plants, found throughout tropical and temperate regions of the world, have been given all sorts of common names alluding to the string-like appearance of the plants: angel hair, devil's hair, devil's ringlet, goldthread, hairweed, lady's laces, strangleweed, witch's hair and many others. These plants have very thin, sting-like twining stems that appear to be leafl ess. They do have leaves, but they are reduced to tiny scales that are barely visible. Most species have very low levels of chlorophyll, so are not green. Because of the lack of chlorophyll most produce very little food on their own, and therefore depend on their host plants for nutrition. The plants cover their host plants in a spreading, tangled mass of intertwined stems, especially where growing in full sun (twining and attachment is greatly reduced in shaded areas).

Orange strands of dodder nearly cover a native burro bush, Ambrosia dumosa, in Anza Borrego Desert State Park, California.

Dodder species are variable in the habitat they are found in naturally and the number of different host species they can infect, with some restricted to just a few host plant species and others able to infect a wide range of hosts. A wide variety of herbaceous and small woody plants can be parasitized by dodder, including many agricultural crops such as alfalfa, asparagus, carrots , cranberries, onions, and potatoes, as well as many ornamental plants, including chrysanthemum, dahlia, helenium, impatiens, English ivy, periwinkle, petunia and trumpet vine, and many weed species including fi eld bindweed (Convolvulus arvensis), lambsquarters (Chenopodium album), and pigweed (Amaranthus species). Dodder can weaken or kill plants and reduce crops yields. The impact on the host plant varies considerably dependin on the species of dodder, the growth stage and condition of the host plant, and the time of infection. Infected plants are also more susceptible to diseases and insect problems.

The thin, string-like stems intertwine over a plant.

Dodder strands on jewelweed, Impatiens capensis, in Wisconsin.

In temperate areas dodder is an annual that grows from seeds each growing season. After germination at or near the soil surface, the seedlings grow quickly toward a host plant, possibly in response to volatiles released by preferred host plants, coiling around any object it encounters. If a suitable host plant is not reached within 5-10 days the seedling will not survive. Once a suitable host plant is reached, the seedling wraps itself around the plant and produces special structures, modifi ed adventitious roots called haustoria, that are inserted into the host's vascular system. Once the dodder is established on the host plant its original root in the soil dies. The dodder plant then continues to grow rapidly on that host plant, continually making new attachments to the host and eventually covering it – or spreading to adjacent plants – until the dodder plant is killed by frost. Since they have to re-establish from seedlings each year, species in temperate areas are only found in relatively low vegetation, whereas species in tropical areas, where plants can grow continuously, may reach high into the canopy of shrubs and trees.

Dodder grows on host plants, wrapping around its stems and leaves, and attaching to its vascular system.

Tiny 4 or 5-parted white, yellow or pink bell-shaped flowers are usually borne in clusters (occasionally singly) in late spring to early fall, depending on the species. These are followed by small rounded fruits or seed capsules the same color as the stems that contain 2-3 seeds. Dodder is a prolific seed producer; a single plant can produce thousands of seeds. The small seeds have a hard, rough seed coat that enables them to survive in the soil up to 20 years or more, depending on the species and environmental conditions. Dodder seeds are spread primarily through the movement of soil and equipment by humans, or in infested plant material or as a crop seed contaminant, and for some species by water.

The small fl owers are produced in clusters (L and LC), with 4 or 5 petals of white or other colors, depending on the species (RC) and are followed by rounded fruits.

Dodder is not commonly found in gardens or ornamental landscapes, but if observed it should be removed immediately as dodder is difficult to control once introduced. Effective management requires a systematic approach combining several control methods over several years focused on reducing the current population, preventing seed production, and dealing with new seedlings in subsequent years. Any seedlings not yet attached to a host plant should be pulled (they are generally difficult to find, however). Once attached to a host plant, complete eradication from the host plant is usually not possible as dodder can grow back from haustoria embedded in the plant. Pruning is of little benefit unless only one or two branches are affected and can be removed without destroying or disfi guring the entire plant. So unless the plant can be pruned signifi cantly lower than the dodder it is best to remove the entire host plant or kill both the dodder and host plant with a non-selective herbicide, such as glyphosate. If the dodder plants have already set seed, burn them (if allowed locally) or dispose of them in the trash. In areas that have been infested previously, the area should be monitored closely for the presence of seedlings and pull them as soon as they are seen or the soil can be treated with a preemergent herbicide before seeds germinate in spring. Control any weeds that could serve as hosts. If the area is planted annually, choose plants that are not susceptible to attack by dodder, such as ornamental grasses and other monocots, including lilies.

In agricultural crops, dodder is managed by rotation to non-host crops, use of dodder-free seed, cleaning equipment between fields, control of weed hosts, and chemical control in infested areas.

Dodder growing on nasturtium.

There are many native species of dodder in Wisconsin, including common or swamp dodder (Cuscuta gronovii (= C. umbrosa) that is potentially invasive, but not all dodder species are pests that need to be controlled. There are several species of native dodders in Wisconsin that are Wisconsin Special Concern plants:

- Hazel dodder (Cuscuta coryli) is found in sandy lake margins, mud flats and moist fields. It is often found on hazelnut (Corylus spp.), as well as reported infecting aster, common milkweed (Asclepias syriaca), Mondarda, sunfl ower (Helianthus sp.), Canada goldenrod (Solidago canadensis), and members of the mint family (Lamiaceae) and many other plants.
- Rope dodder (Cuscuta glomerata) is found in a variety of moist habitats, and blooms from July through September. It can be distinguished from other Cuscuta by having the infl orescences densely bunched together giving the plant a thick, rope-like appearance. It is reported as having a preference for members of the Aster family, often on sunflower, Canada goldenrod, and spotted water hemlock or spotted cowbane (Cicuta maculata).
- Field or five-angled dodder (Cuscuta pentagona = C. arvensis, C. campestris) is found along fields and roadsides.
- Knotweed or smartweed dodder (Cuscuta polygonorum) is also found in moist habitats including wet prairies, coastal plain marshes, and the margins of commercial cranberry bogs. It is distinguished from other Cuscuta by itsvery short styles and its host plants including Bidens spp., Polygonum spp., Panicum capillare, Setaria glauca, Rorippa islandica, and Hypericum boreale.

Rope dodder, Cuscuta glomerata.

Most species of dodder are on the federal noxious weed list, except some native species and a few widespread non-native species, but all dodders, including the natives, require federal permits for importation or transportation of seed.

Orobanche Aegyptiaca

Orobanche aegyptiaca, Egyptian broomrape, is a plant which is an obligate holoparasite from the family Orobanchaceae with a complex lifecycle. This parasite is most common in the Middle East and has a wide host range including many economically important crops.

Selective control of Egyptian broomrape is extremely difficult because the close association between host crop and parasite limits the use of most mechanical and herbicidal approaches.

Disease Cycle

A single Egyptian broomrape plant is capable of producing hundreds of thousands of extremely small (0.15-0.5 mm long) seeds. These seeds, dispersed by the wind, animals, or by more artificial means such as farm machinery, survive in the soil and have the ability to remain viable in the soil for more than 15 years. When the seeds are dropped, they are dormant. They require after-ripening, or further ripening after the seed has fallen from the plant, which is completed by the time the dry season has ended. As the rainy season begins, the seeds imbibe water and move into a conditioning phase which lasts 1–3 weeks, depending on whether the temperature remains optimal, between 60 and 70 °F. At this point, the seeds only need a chemical signal from the roots of a host plant to germinate. The seeds can remain in this conditioned state for several months. If the seed never receives the chemical signal to germinate and dries out, it returns to its dormant state, ready to be reconditioned when the next rainy season begins.

When a conditioned seed receives a chemical signal from a host plant, it germinates, forming a germ tube which grows toward the host. From this germ tube, it grows a haustorium, which is the part of the broomrape which attaches the parasite to the host and grows into the host's vascular system. The parasite is then able to draw out water and nutrients from its host. At this point, O. aegyptiaca grows rapidly, developing above-ground flower stalks. The flowers develop between February and May. A medium-sized plant can produce more than 400 flowers, which each produce around 500 seeds. The complete lifecycle of this parasite, if completed in one season, lasts 10 to 15 weeks.

Hosts and Symptoms

The host range for Egyptian broomrape is fairly wide, including many broadleaf vegetables, field crops, and some ornamentals. Tomato, potato, tobacco, eggplant, peppers, peas, carrot, celery, mustard, spinach, and chrysanthemum are among the susceptible plants. In areas such as southern Russia, melons are also potential hosts. Some plant species have been reported as being hosts for O. aegyptiaca that, in reality, are not hosts. Because the physical attachment of the parasite occurs below ground, checking that the plant's roots are connected to the weed is essential.

The most obvious sign of O. aegyptiaca parasitism is the visible parasitic plant near the base of the plant being attacked. Egyptian broomrape is a very short plant with purple flowers and a purple to brown stem. Plants that have been parasitized by this weed will show evidence of water stress, leaf yellowing, and stunted growth.

Environment

O. aegyptiaca is adapted to soils with a high pH, like those in the Middle East where it is native, and requires high temperatures for ideal germination and growth. Because O. aegyptiaca is an obligate parasite it is found only in association with the crops it attacks, especially irrigated crops.

Management

Three main types of control exist for Orobanche aegyptiaca: chemical, cultural, and biological. Chemical control can be achieved by soil fumigation with methyl bromide. This method is effective yet is rarely used because only a shallow layer of the soil is affected, it is a costly treatment, and methyl bromide poses environmental concerns. Another form of chemical control is soil solarization, which is a solar heating of the soil by placing clear polyethylene sheets over moist soil. This kills the majority of seeds that are viable and induces secondary dormancy in the rest. Lastly, sulfonylurea and imidazolinone herbicides have been shown in many studies to be an effective form of chemical control for O. aegyptiaca with the efficacy depending on the method of application, the species of crop, and the timing of the application. One study has shown application times between 14 and 42 days after planting to be ideal for controlling Egyptian broomrape.

Some forms of cultural control of O. aegyptiaca include sanitation efforts to help prevent the movement of seeds, individual picking of weeds by hand, and avoidance by changing the sowing time or not growing a host plant. Reducing spread of infected soil by farm machinery, avoidance of grazing on infected plants, and staying away from the use of hay produced from Orobanche-infested plants are all methods to help prevent the movement of seeds. The individual picking of weeds by hand is very important, as the plant is able continue living as only a stem and produce a flower that can spread seeds even while not connected to the host. Any avoidance strategy must be carried out with a plant that cannot be parasitized by O. aegyptiaca and could take up to 20 years to complete, as this is how long the plant's seeds can remain viable in the soil. Also, because seeds can safely pass through an animal's digestive system, ensuring manure is not contaminated is important. Another option for cultural control is the use of trap crops or catch crops. Trap crops promote the germination of O. aegyptiaca seeds, but do not allow parasitism; these include flax, mung bean, maize, and sorghum. Catch crops allow parasitism, but are destroyed before the parasitic plants flower, so the broomrape seeds cannot be produced and dispersed. Finally, limited success has been seen with host plant resistance in sunflower, faba bean, lentil, and tomato. Resistant cultivars of sunflower were bred in Russia and Spain, but quickly lost their host-plant resistance due to selection towards more aggressive biotypes of the plant that had adapted to the new cultivars. In the resistant sunflower cultivars, germination was still induced, but the plant's germ tube never develops after penetration by the haustorium. Macromolecules are transferred from the host to the parasite, so one method for resistance is the introduction of an antimicrobial gene from the flesh fly into the host genome. This allows for its protein product to be transferred to the parasite and work as a means of control.

Most forms of biological control are still being developed and studied, although three species are potential forms of control: Fusarium oxysporum, F. solani, and Sclerotinia sclerotiorum. F. oxysporum has been shown to successfully control O. aegyptica in sunflower and tobacco, while S. sclerotiorum has been shown to cause wet rot in the plant without having any negative impact on the host. Field experiments carried out in Hungary showed promise for control with some "F. solani" isolates. The "Fusarium" species usually work through phytotoxins that help them to conquer O. aegyptica 's defenses and establish themselves.

Importance

Because Egyptian broomrape can cause severe damage to economic crops and ornamentals, it is important to understand the plant as a parasite and to fully comprehend its life cycle. By understanding its lifecycle, management is possible.

In Israel, the production of processing tomatoes is in danger as a result of high crop yield losses due to heavy infestations of O. aegyptiaca. Approximately $1.3 to 2.6 billion have been lost in annual food crop losses in the regions near the Mediterranean, Northern Africa, and Asia. Egyptian broomrape and branched broomrape (O. ramosa) have infested an annual average of 2.6 million ha of crops from the Solanaceae family, including tomatoes, potatoes, and eggplant.

Orobanche Minor

Orobanche minor, (hellroot, common broomrape, lesser broomrape, small broomrape or clover broomrape), is a holoparasitic angiosperm belonging to the genus Orobanche; a genus of about 150 non-photosynthetic plants that parasitize other autotrophic plants.

Characteristics and Growth Requirements

O. minor grows to 0.5 m (1 ft 8 in) and is a perennial. The flowers are hermaphrodite. Common broomrape grows in a wide variety of soils, namely moist, light (sandy), medium (loamy) and heavy (clay) soils that are acid, neutral or basic. It can grow in semi-shade or in full sunlight.

The species appears in a wide range of colours from red-brown, yellow-brown to purple. Yellow specimens are also not uncommon and it is this extreme variability that makes identification on the basis of size or colour uncertain. It is parasitic on various members of the pea (Fabaceae) and daisy (Asteraceae) families. Although widespread, its appearance is sporadic; despite this, it can occur in vast colonies from time to time. The main flowering season in the northern hemisphere is from May until the end of August and from August to January in the southern hemisphere. The species has efficient seed dispersal and is largely inbreeding so that populations preferentially parasitizing a particular species which has its own clear ecological preferences may become effectively isolated and eventually may produce distinct taxa.

The plants are attached to their host by means of haustoria, which transfer nutrients from the host to the parasite. Only the hemiparasitic species possess an additional extensive root system. The root system is reduced as its function is mainly anchorage of the plant.

Distribution

Common broomrape is one of the most widespread species, and is native to Southern Europe, but has been widely introduced elsewhere, for example in the United States. In New Zealand it is the only species of the genus present where it is regarded as an agricultural pest. In the United Kingdom it is widely recorded in southern England, less common in Wales, rarely recorded in lowland Scotland and absent from the Highlands and outer islands.

Taxonomy

Phylogenetic analyses have placed this species in the taxonomically difficult Minores species complex. Four infraspecific taxa of Common broomrape are currently recognised in the United Kingdom: O. minor var. minor, O. minor var. flava, O. minor var. compositarum and O. minor subsp. maritima. Chromosome No.: 2n = 38. The genetic structure of populations of O. minor are under investigation using molecular markers and DNA sequencing to help resolve the taxonomic and nomenclatural problems that have historically been linked with this species.

Hosts and Speciation

The common broomrape is highly generalist in its host range, and can infect hundreds of species in families from the Ranunculaceae to the Poaceae (=Gramineae) but with a clear preference for the Fabaceae (=Leguminosae) and Asteraceae (=Compositae). However races occurring on different species of host are genetically divergent and physiologically adapted to their local hosts, and may therefore be in a state of incipient speciation. Urgent conservation efforts are required as the survival of some intraspecific taxa is very uncertain.

Entirely yellow O. minor colony.

Rhinanthus Minor – Yellow Rattle

Yellow rattle is a grassland annual with yellow, two lipped flowers, the upper lip with two white or purple teeth. Behind the flower the joined sepals inflate to form a green bladder sometimes tinged with red. Flowering occurs from May through to September followed by a seed capsule which rattles within the now brown and papery bladder. The majority of seed ripens in mid to late July and the rattling of the fruit within the bladder was said to indicate that the meadow was ready to be cut for hay, hence its other common name of hay rattle.

Type	Seeds per gram	Origin	Ordering
Meadow annual	300	Somerset, Kent, Caernarvonshire	Order this species

Habitat Information

Yellow rattle is an annual root-hemiparasite of moderate to low fertility grasslands particularly meadows but also roadsides, dunes, lightly grazed pastures and dry areas within fens. It parasitizes the roots of a wide range of meadow plants especially grasses and legumes and whilst capable of carrying out its own photosynthesis it is dependent upon these hosts for additional supplies of carbohydrates and minerals. By drawing nutrients from surrounding vegetation it impedes their growth and helps maintain an open sward structure. Yellow rattle is however intolerant of shade and on fertile sites where coarse grasses are present or on grasslands that are never cut or grazed it tends to get shaded out. The seed, which is set from late June onward, is winged and dispersed a short distance by the wind. Seed requires a period of chilling to break dormancy and germinates in early spring. There is no persistent seed bank in the soil.

Yellow rattle populations tend to fluctuate in meadows and 'move about' from year to year as a reflection of the balance of health of the Yellow rattle plants and their host plants in any one patch.

Growing Information

Yellow rattle seed must be sown in autumn as it needs a prolonged period of chilling through the winter to trigger its germination the following spring. Yellow rattle can be sown as part of a mixture, or on its own into established grassland prepared as below.

The most suitable sites for Yellow rattle will be grassland of low to medium fertility and with a balanced sward which is not dominated by coarse vigorous grasses. Prepare for sowing by cutting the grass very short (25mm) or by grazing very hard and open up sites for germination by harrowing, raking or lightly discing, aiming to create up to 50% bare soil. Broadcast the seed on to the prepared surface.

Traditional meadow management based around a late July hay cut provides Yellow rattle the best opportunity to set seed, and for its seed to scatter during the process of haymaking.

Sow Yellow rattle at a rate of between 0.1g and 1g per square metre. The rate chosen is most usually a compromise between the desire to rapidly attain a plant population density sufficient to suppress grass and the available budget.

Viscum Album

Viscum albun, also known as mistletoe, is a semi-parasite, which grows mainly on deciduous trees and spread in the temperate zones of Europe and Asia. As a semi-parasite it lives on trees and cuts off the "host´s" supply of water and nutrients. The mistletoe produces energy-rich organic compounds through the photosynthesis of its own green leaves throughout the year. Thus, the host tree is not permanently damaged by the mistletoe. If the deciduous trees have shed their leaves in autumn, you can see the mistletoe on poplar, birch, willow and other deciduous trees. They are spherical subshrubs with a diameter of about 1m wide, reminiscent of a bird´s nest. Green-brown, dichotomously branched twigs spring from a short trunk. The leaves are leathery, with entire margins, lanceolate to broadly ligulate. At the end of each forked part there is a flower-bearing top. The flowers themselves are rather inconspicuous. What is remarkable are the pea-sized white berries with their slimy sticky content that form from the female flowers. Flowering time is February to May, the berries ripen in late autumn. According to an old custom the green mistletoe with the white berries are attached on the doors of the houses at Christmas time to ward off demons.

The Latin name Viscum album may make reference to the fact, that the so-called bird lime was formerly made from the white berries. It was used to coat limed twigs and thus began the coveted meal of songbirds, especially the Mistle Thrush (Turdus viscivorus). The berries are eaten by the thrushes, the seeds pass undigested through the bird´s intestine and area disseminated in their droppings.

Applications

Segment therapy for degenerative inflammatory joint diseases by producing cuti-visceral reflexes after calming local inflammation through intradermal injections; for palliative therapy in terms of a nonspecific stimulation therapy for malignant tumours. These applications were approved by the Commission E for mistletoe; mistletoe stems and mistletoe berries were given negative monographs (unfavourable benefit-risk ratio). The treatment monographs developed by Commission E (phytotherapy), C (anthroposophy) and D (homeopathy) on mistletoe were during the years: 1984 to 1994. Even after this time the mistletoe has been and is well researched; its immune-stimulating anti-cancer effect has been studied, focusing on the mistletoe lectins contained in the mistletoe. They influence the release of cytokines, which act as mediators for important processes of the immune system and so can also be used to fight tumour cells. Cytokines however, have two faces, because they can stimulate the division of tumour cells and thus by improving the immune status accelerate the growth of the tumour. Clinical trials with cancer patients who were treated with a standardised phytotherapeutic with mistletoe lectin content have so far have not shown any remarkable effects. However, there was improvement in the quality of life, appetite, mood and the general condition and performance of the patients. The evidence base has led to the approval of mistletoe injection preparations "for the palliative treatment in terms of a nonspecific stimulation therapy for malignant tumours" (allopathic phytotherapy). Treatment is in the hands of experienced doctors. Another approach is the background of anthroposophical cancer therapy with anthroposophical mistletoe preparations. It goes back to Rudolf Steiner, who introduced the use of mistletoe extracts in cancer therapy in 1921. He declared the special spirituality of anthroposophical mistletoe preparations as part of his metaphysical and esoteric, dogmatic pharmacology was a causal factor in healing. Such preparations are designated in pharmaceutical legislation as "Special treatment directions" and may according to the findings of the Commission C (formerly responsible for anthroposophic medicinal products) be used "for the treatment of malignant and benign tumour diseases, malignant diseases and associated disorders of the blood-forming organs and stimulate bone marrow growth and to prevent tumour recurrence after surgery." This form of treatment can only be performed by trained doctors. The scientific evidence is pending.

Traditional Use

Mistletoe is also traditionally used in combination with other drugs to support the cardiovascular function.

Medicinal Herbal Preparations in Finished Drug Products

Phytotherapy for the palliative treatment of tumours:

- Aqueous extract in injection solutions.
- Fluid extract in injection solutions.

Phytotherapy to Support the Circulatory Function

- Cut Drug to prepare tea.
- Powder in tablets.
- Dried extract in pills.
- Tincture and alcoholic extracts in drops.
- Mother Tincture in drops.

Anthroposophic Medicine for Cancer Therapy

- Fresh vegetable juices from different mistletoe host trees in injection solutions.
- Aqueous extracts from fermented apple tree mistletoe in injection solutions.

Epifagus

Epifagus virginiana — commonly called beech drops (or beech-drops, or beechdrops) — is an obligate parasitic plant which grows and subsists on the roots of American beech. It is a member of the broomrape family. Epifagus is monotypic—containing only E. virginiana. The name Epifagus derives from Greek "epi" meaning "on" or "upon", and "Fagus" which is the genus name of beech.

Beechdrops is an annual plant native to eastern North America. It entirely lacks chlorophyll and produces many brown stems up to 30 cm tall on which it bears small white and purple flowers that appear in July through October. The flowers have tubular, zygomorphic, corollas ~8mm long containing a single style and four stamens. The dried flower stalks will persist throughout the winter. The flowers on the lower parts of the plant are cleistogamous (self-pollinating) while the tops of the stems have chasmogamous (cross-pollinating) flowers which may be sterile.

Beechdrops does contain very small alternate, scale like leaves, which are a vestigial structure from a common ancestor which was photosynthetic.

Host and Symptoms

Epifagus virginiana, also known as Beechdrops, is an obligate parasite to Fagus grandifolia, Beech trees. It has been found on Maple trees but it is believed this is a case of mistaken identity. E. virginiana grows off of the roots of its host but is not known to cause significant harm to the Beech tree. Beechdrops grow on shallow roots at varying distances form the trunk of F. grandifolia. The roots of the host release a chemical that trigger the germination of E. virginiana. It is believed that the older the host tree, the more this chemical is released. The parasite develops a hastorium that grows into Beech roots to draw nutrients from its host, as Beechdrops do not photosynthesis. E. virginiana does not tolerate disturbances in its connection to F. grandifolia. As E. virginiana develops a tuber the hastorium diminishes to the point that when E. virginiana has reached maturity there is no visible hastorium, instead the parasite and host are connected through their roots and the Beechdrops' tuber.

Life Cycle

E. virginiana germinates when a chemical signal is released from the beech tree's roots. During early stages of development, beechdrops live independently from its host instead relying on nutrients from the seed. It may take several years for an above ground structure to form. These early stages E. virginiana are a few millimeters in size and butter yellow. Beechdrops then develop a hastorium that grows into the host's roots. At maturity the hastorium degrades and a tuber develops. During late summer and early fall E. virginiana flowers. Beechdrops produces two types of flowers: chasmogamous and cleistogamous. Chasmogamous flowers are cross pollinated flowers that grow at the top of the plant and are sometimes sterile. Cleistogamous flowers are self- fertile, these flowers grow at the base of the plant. Seeds from E. virginiana are small and are dispersed by rainwater. Between dispersal and germination, seeds experience a cellular change, the embryo changes colors and its cells develop granules.

Environment

E. virginiana's host environment is mainly temperate forests in the Midwest of the U.S. Beechdrops are found in Indiana, Ohio, and Michigan in significant numbers. They have been documented along the East Coast from Maine to South Carolina. In these environments E. virginiana are pollinated by P. imparis, the winter ant. This insect pollination is required for pollination in the chasmogamous flowers. E. virginiana relies upon ant pollination to produce cross pollinated seeds along with self-fertilized seed. It is not believed that ants pollinate the self-fertile cleistogamous flowers. Beechdrops are used to monitor forest health because of their dependence on their host and the sensitivity to its environment. E. virginiana's host, Beech trees, can advance in a forest faster than E. virginiana is able to. The presence of F. grandifolia is used to predict when E. virginiana will arrive in an area. E. virginiana is not a tree health concern but the lack of its presence is a sign that forest health is declining.

Nuytsia

Nuytsia floribunda is a hemiparasitic tree found in Western Australia. The species is known locally as moodjar and, more recently, the Christmas tree, as the display of intensely bright flowers during the austral summer coincides with the Christmas season.

The habit of the species is a tree up to 10 metres (33 ft) high, or as a shrub. The rough bark is grey-brown. Flowers are a vivid yellow-orange, appearing between October and January. The inflorescence on each flowering stem may be up to one metre in length.

It is a root hemiparasite, is photosynthetic and mainly obtains water and mineral nutrients from its hosts. The haustoria arising from the roots of Nuytsia attach themselves to roots of many near-by plants and draw water and therefore nutrients from them. Almost all species are susceptible to attack; haustoria have even been found attached to underground cables. In natural settings Nuytsia withdraws relatively little from each individual host, but is attached to so many other plants that the benefit to this hemiparasitic tree is likely to be considerable. Roots and rhizomes extend out and may sucker to form new branches that give the appearance of a grove of trees. A network of fine and fragile roots arise from these larger underground parts, forming haustoria where they meet the roots of other species.

The trunk is reported as up to 1.2 metres in diameter, with multiple layers of wood and bark that allow the tree to withstand fire. An unusual characteristic of the seedlings is the four to six cotyledons rather than two.

Fruit

Taxonomy

Recognised in early descriptions as a species of the family Loranthaceae, a group almost entirely represented by epiphytic mistletoes, allied to the order Santalales. The current treatment of Nuytsia floribunda is as a monotypic genus, Nuytsia. This species was seen as a sister taxon to Atkinsonia ligustrina (Loranthaceae) in the description Nuytsia ligustrina by Allan Cunningham in 1817, before its later arrangement to a separate genus, Atkinsonia, which also describes a single species of Loranthaceae.

The first description of Nuytsia floribunda was published by Jacques Labillardière in his 1805 work Novae Hollandiae Plantarum Specimen, as Loranthus floribundus, the specific epithet describing the profuse flowers he would have observed at Esperance. The botanist Robert Brown published a remark on the species in 1831, giving a new genus name without a formal description. The description was published by George Don using Brown's name Nuytsia, an epithet that commemorates the seventeenth-century Dutch explorer and colonial official Pieter Nuyts.

The Noongar names for the plant are moojar, moojerool, munjah and mutyal. The regional name of 'Christmas tree' has been applied since the nineteenth century. James Drummond noted in 1843 that the colonists at the Swan River referred to this species as the 'fire-tree', given for the resemblance of the inflorescence to a fire in its habitat. William Milligan reported that the first colonists used the term 'cabbage tree', a reference to its fragile, white and spongey branches that can be snapped off as easily as the stalks of the European vegetable, cabbage.

Distribution and Habitat

Nuytsia floribunda is well known in Southwest Australia, especially due to the appearance of abundant flowers in summer which is a spectacular display. Although Nuytsia seeds germinate readily and seedlings are easy to grow for a year or two, cultivation of the species to maturity is regarded as difficult, with little success outside of its native habitat. It appears on a variety of soil types throughout Southwest Australia, the distribution of the species extends to the east of the Esperance Plain and to the north on the Geraldton Sandplains.

The species was once common across the Swan Coastal Plain, now mostly cleared around Perth with changes in land use since colonisation.

These beautiful flowers are parasites and flower here every year for Christmas. Nuytsia floribunda.

Christmas tree, Nuytsia floribunda, in bud. Kensington Bushland, Perth, WA 6151.

Uses of Nuytsia

The Nyungar people made use of the species during the season Kambarang, around October to early December, obtaining bark to make shields. The gum that exudes from the wound can be collected later, it is sweet and eaten raw. Moodjar (or Muja) is regarded as a protected tree by the Nyungar peoples of Southwest Australia, the species is noted as being incorporated into rituals and having a conservation status that forbids their destruction. The plant is venerated by some who learn that it should not be sat beneath or its flowers, leaves or branches touched or taken. The sugary gum is consumed in modest quantities, and children are warned of overindulgence with the story of a monstrous, invulnerable and inescapable nocturnal being who cry of 'Nhervalong' could be heard as it collects the gum on which it subsists. Seasonal use by other groups of the roots is also known, Ethel Hassell at Jerramungup was invited to taste some being harvested by a group of women, reporting the taste of the flesh beneath an easily removed skin as sweet, brittle and somewhat watery.

The wood of the tree was noted as commercially useless, and that cultivation and transplanting difficult or impractical for gardens or landscaping by the state's conservator Charles Lane Poole, deeming the showy display of flowers to be its only desirable quality. Poole also notes the 'gum' that is exuded from the tree may be useful. The artist Marianne North featured the tree in an illustration titled 'Study of the West Australian Flame-tree or Fire-tree' now held at Royal Botanic Gardens, Kew, and recorded her effusive impressions, "I shall never forget one plain we came to, entirely surrounded by the nuytsia or mistletoe trees, in a full blaze of bloom. It looked like a bush-fire without smoke. The trees are, many of them, as big as average oaks in our hedgerows at home" The early settler George Fletcher Moore described this parasitic tree as another "anomaly in this land of contradictions".

Ecology

The flowers produce large amounts of pollen and nectar that is consumed by insects. The parasitic relationship of Nuytsia floribunda on host species was poorly understood until an investigation by the botanist D. A. Herbert was presented to the Royal Society of Western Australia, contradicting the assumption that proximity to other tree species of banksia and jarrah, Eucalyptus marginata, was a requirement. Herbert discovered the extended network of filaments that embrace the roots of many other species, explaining the persistence of trees conserved in agricultural land cleared for introduced crops.

Nuytsia floribunda was once common and well known on the coastal plain around Perth, often remaining in remnant bushland and gardens, becoming more scarce as the extent and density of urban development increased.

Rafflesia Arnoldii

Rafflesia arnoldii, commonly called the corpse lily, is a species of flowering plant in the parasitic genus Rafflesia. It is noted for producing the largest individual flower on Earth. It has a very strong and unpleasant odour of decaying flesh, earning it the nickname "corpse flower". It is endemic to the rainforests of Sumatra and possibly Borneo. Although there are some plants with larger flowering organs like the titan arum (Amorphophallus titanum) and talipot palm (Corypha umbraculifera), those are technically clusters of many flowers.

Rafflesia arnoldii (Indonesian: padma raksasa), also called Kerubut (Devil's Betelnut Box), is one of the three national flowers in Indonesia, the other two being the white jasmine and moon orchid. It was officially recognized as a national "rare flower" (Indonesian: puspa langka).

The flower of Rafflesia arnoldii grows to a diameter of around one meter (3.3 feet), but the greatest measurement from a reliable source is 105 centimeters (3.44 feet) for one at Palupah Nature Reserve near Bukittinggi, Sumatra measured by Prof. Syabuddin of Andalas University. R. arnoldi weighs up to 11 kilograms (24 lb). These flowers emerge from very large, cabbage-like, maroon or magenta buds typically about 30 cm (12 in) wide, but the largest (and the largest flower bud ever recorded) found at Mount Sago, Sumatra in May 1956 was 43 cm (17 in) in diameter. It lives as a parasite on several vines of the genus Tetrastigma, which grow only in primary (undisturbed) rainforests. Rafflesia lacks any observable leaves, stems or even roots, yet is still considered a vascular plant. Similar to fungi, individuals grow as thread-like strands of tissue completely embedded within and in intimate contact with surrounding host cells from which nutrients and water are obtained. This plant produces no leaves, stems or roots and does not have chlorophyll. It can only be seen outside the host plant when it is ready to reproduce. Perhaps the only part of Rafflesia that is identifiable as distinctly plant-like are the flowers, though even these are unusual since they attain massive proportions, have a reddish-brown coloration, and stink of rotting flesh. This scent attracts insects such as flies which then pollinate the rare plant.

Specimen of Rafflesia arnoldii at the Kyoto Botanical Garden.

The first European botanist to find a specimen of a Rafflesia was the French explorer Louis Auguste Deschamps. He was a member of a French scientific expedition to Asia and the Pacific. During the expedition, he spent three years on Java, where, in 1797, he collected a specimen of what is now known as R. patma. During the return voyage in 1798, his ship was taken by the British, with whom France was at war, and all his papers and notes were confiscated. They did not see the light of day until 1954, when they were rediscovered in the Natural History Museum, London.

The British botanist Joseph Arnold and the statesman Sir Stamford Raffles collected a specimen of another Rafflesia species found by a Malay servant in Sumatra (Bengkulu) in 1818. Arnold contracted a fever and died soon after the discovery. Lady Raffles, who had also been present when the specimen was collected, finished the colour drawing that Arnold had started of the plant, and it was sent to Joseph Banks, along with the preserved material. Banks passed all the materials on to Robert Brown of the British Museum and Kew's resident botanical artist Franz Bauer.

William Jack, who was Arnold's successor in Sumatra, being aware that Deschamps, despite his loss of notes, could formally publish a name for the newly discovered genus at any moment, rushed to draft a description to ensure the credit went to a British botanist. This draft description was held in readiness, in case there was word that the French were about to publish, whilst waiting for the British Museum to produce a better-prepared version.

The generic name Rafflesia, given in honour of Raffles and proposed by Brown, who had originally wanted to call it Arnoldii after Arnold, was validated by S. F. Gray in his report of the June 1820 meeting of the Linnean Society of London, as published in the Annals of Philosophy in September that year. The species Rafflesia arnoldii was officially described for the first time in 1821 by Brown, so that Arnold was commemorated after all.

Reproduction

Rafflesia arnoldii is rare and fairly hard to locate. It is especially difficult to locate the flower in forests, as the buds take many months to develop and the flower lasts for just a few days. The flowers are unisexual and thus proximity of male and female flowers is vital for successful pollination. These factors make successful pollination a rare event.

When Rafflesia is ready to reproduce, a tiny bud forms outside the root or stem of its host and develops over a period of a year. The cabbage-like head that develops eventually opens to reveal the flower. The stigma or stamen are attached to a spiked disk inside the flower. A foul smell of rotting meat attracts flies and beetles to pollinate. To pollinate successfully, the flies and beetles must visit both the male and female plants, in that order. The fruit produced are round lots filled with smooth flesh including many thousands of hard-coated seeds that are eaten and spread by treeshrews.

Mature Rafflesia arnoldii flower with buds.

Threats

Many areas where Rafflesia arnoldii grows have become popular tourist attractions. While such tourism has helped locals to conserve the species, the increased human activity has disturbed and reduced flower bud production at many locations.

Conservation Status

How many of these plants still survive is unknown, but as the remaining primary forests of Borneo and Sumatra disappear, it can be assumed that their numbers are dwindling. Many are known to be nearing extinction. Some environmentalists are developing ways to recreate the species' environment in an effort to stimulate their recovery. This has proved unsuccessful so far. Steps are also being taken to conserve the forests of Sumatra and Borneo. To help counter the over-collection of this rare plant, residents that have Rafflesia on their private property are encouraged to save the flowers and charge a small fee to see them.

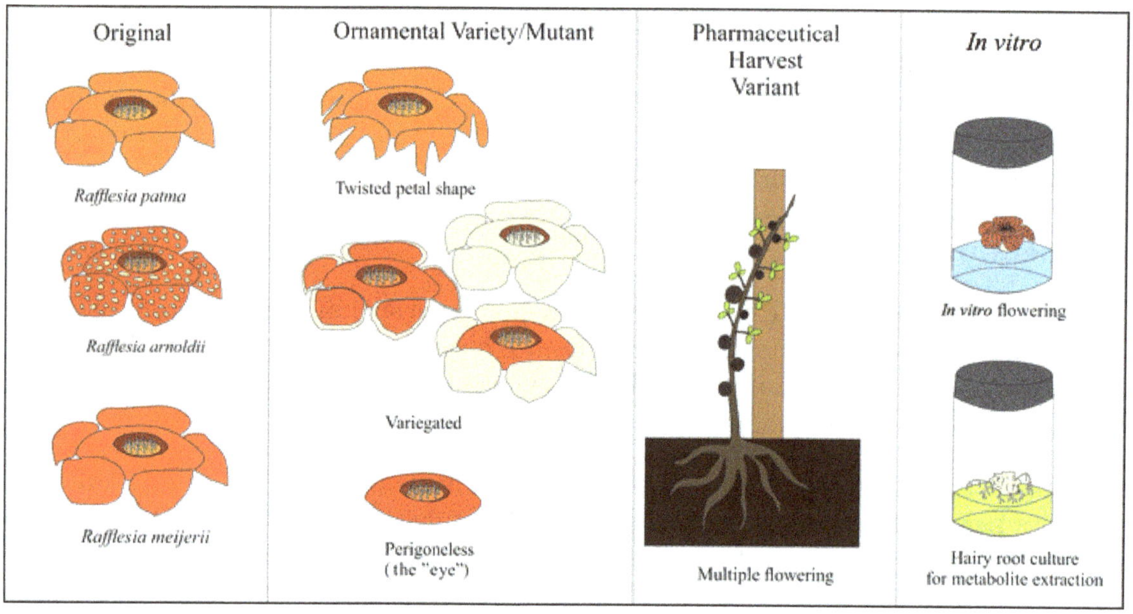

Orobanche Uniflora

Orobanche uniflora, commonly known as one-flowered broomrape, one-flowered cancer root, ghost pipe or naked broomrape, is an annual parasitic herbaceous plant. The "broom" might refer to the plant's shape while "rape" might refer to the lumps that are on its roots. It is native to much of North America, where it is a parasitic plant, tapping nutrients from many other species of plants, including those in the families Asteraceae and Saxifragaceae and in the genus Sedum. The name "orobanche" can be translated to "vetch-strangler" and "uniflora" can be translated to "single-flower".

Orabanche uniflora is 3.5 centimetres (1.4 in) long. The corolla is violet to yellow-brown, two-lipped, finely fringed with five similar lobes. The species has hairless bracts and purple or white flowers with five petals. The main stem is under the ground, with only the pedicels being seen and each pedicel containing only one flower. The stems are grayish tan. No leaves are on the plant or offshoot from it. It reproduces from its seeds, which are produced from fruit that has two sections. Many seeds are produced from the plant's fruit.

The plant is parasitic, feeding off of other plants' root systems, often on the flowering plant genus Sedum which are also known as stonecrops. It does not produce chlorophyll, rather gaining its nutrients, such as carbohydrates, from other plant species. Due to not producing chlorophyll, the plant does not produce any green portions. It is not agreed upon by botanists whether the plant's flowers are pollinated by insects or if they are pollinated by the plant itself. The life expectancy of the species is unknown.

The species closely resembles Orobanche fasciculata. The two can be differentiated by their bracts and number of flowers. O. uniflora's bracts are hairless while O. fasciculata's bracts are hairy. The flowers of O. uniflora have one to three flowers on each stem branch while O. fasciculata has three to ten flowers per stem branch. O. fasciculata has flowers that are pinkish to creamy-white, unlike O. uniflora which is tinged with violet.

Habitat

The plant can be found in woods, thickets, and mountains as well as by stream banks, and is widespread in much of North America. Unlike other species in the genus Orobanche, Oranbanche uniflora is largely a species that can be commonly found in forests. The species has been found surviving in both the sunlight and shaded areas, as well as in a variety of different types of soil. Its habitat is restricted to places where there are plenty of host plants to get nutrients from. It might be difficult for this species to survive in an area where its host plants are sparsely found.

A specimen is contained within the botany department of the Smithsonian National Museum of Natural History and another specimen has been photographed at Hells Canyon National Recreation Area. It is considered rare or vulnerable in 17 states and five Canadian provinces. The species was once labeled as being a Special Concern in Minnesota in 1984, later being changed to endangered in 2013. When it was originally listed as a Special Concern in Minnesota, only seven populations were known to exist.

References

- Sedigheh S, Aptin R, Zoheir Y.A, Application Soil Solarization, on the Control of Egyptian Broomrape in Greenhouse, International Journal of Natural and Engineering Sciences 3 (1): 59-64, 2009
- Manen, JF; Habashi, C; Jeanmonod, D; Park, JM; Schneeweiss, GM (2004). "Phylogeny and intraspecific variability of holoparasitic Orobanche (Orobanchaceae) inferred from plastid rbcl sequences". Molecular Phylogenetics and Evolution. 33: 482–500. Doi:10.1016/j.ympev.2004.06.010. PMID 15336681
- "View of A Preliminary Morphological Study of Epifagus virginiana (L.) Bart". Journals.iupui.edu. Retrieved 2018-12-08
- Mistletoe, medicinal-plants: koop-phyto.org, Retrieved 16 August, 2019
- Spellenberg, Richard (2001). National Audubon Society Field Guide to Wildflowers Western Region. Alfred A. Knopf, Inc. P. 652. ISBN 9780375402333
- Ryan, John Charles (2014). "Overland from Perth to Albany: a journey of botanical intrigue". Australian garden history. Australian Garden History Society. 25 (3 (2014:Jan/Feb/Mar.)): 7. ISSN 1033-3673
- Abbate, Campbell (January 2013). "Parasitic Beechdrops (Epifagus virginiana): A Possible Ant-Pollinated Plant". Southeastern Naturalist. Retrieved 8 November 2015

Permissions

All chapters in this book are published with permission under the Creative Commons Attribution Share Alike License or equivalent. Every chapter published in this book has been scrutinized by our experts. Their significance has been extensively debated. The topics covered herein carry significant information for a comprehensive understanding. They may even be implemented as practical applications or may be referred to as a beginning point for further studies.

We would like to thank the editorial team for lending their expertise to make the book truly unique. They have played a crucial role in the development of this book. Without their invaluable contributions this book wouldn't have been possible. They have made vital efforts to compile up to date information on the varied aspects of this subject to make this book a valuable addition to the collection of many professionals and students.

This book was conceptualized with the vision of imparting up-to-date and integrated information in this field. To ensure the same, a matchless editorial board was set up. Every individual on the board went through rigorous rounds of assessment to prove their worth. After which they invested a large part of their time researching and compiling the most relevant data for our readers.

The editorial board has been involved in producing this book since its inception. They have spent rigorous hours researching and exploring the diverse topics which have resulted in the successful publishing of this book. They have passed on their knowledge of decades through this book. To expedite this challenging task, the publisher supported the team at every step. A small team of assistant editors was also appointed to further simplify the editing procedure and attain best results for the readers.

Apart from the editorial board, the designing team has also invested a significant amount of their time in understanding the subject and creating the most relevant covers. They scrutinized every image to scout for the most suitable representation of the subject and create an appropriate cover for the book.

The publishing team has been an ardent support to the editorial, designing and production team. Their endless efforts to recruit the best for this project, has resulted in the accomplishment of this book. They are a veteran in the field of academics and their pool of knowledge is as vast as their experience in printing. Their expertise and guidance has proved useful at every step. Their uncompromising quality standards have made this book an exceptional effort. Their encouragement from time to time has been an inspiration for everyone.

The publisher and the editorial board hope that this book will prove to be a valuable piece of knowledge for students, practitioners and scholars across the globe.

Index